Springer Series in **Materials Science**   3

Edited by Aram Mooradian

Springer Series in **Materials Science**

Editors: Aram Mooradian   Morton B. Panish

Ian W. Boyd

Laser Processing of Thin Films and Microstructures

Oxidation, Deposition and Etching of Insulators

With 77 Figures

Springer-Verlag Berlin Heidelberg New York
London Paris Tokyo

Dr. *Ian W. Boyd*

Department of Electronic and Electrical Engineering,
University College London, Torrington Place, London WCIE 7JE, UK

Series Editors:
Dr. *Aram Mooradian*

Leader of the Quantum Electronics Group, MIT,
Lincoln Laboratory, P.O. Box 73, Lexington, MA 02173, USA

Dr. *Morton B. Panish*

AT&T Bell Laboratories, 600 Mountain Avenue, Murray Hill, NJ 07974, USA

ISBN-13:978-3-642-83138-6   e-ISBN-13:978-3-642-83136-2
DOI: 10.1007/978-3-642-83136-2

Library of Cóngress Cataloging-in-Publication Data. Boyd, Ian W., 1958- Laser processing of thin films and microstructures ; oxidation, deposition, and etching of insulators. (Springer series in materials science ; 3). 1. Lasers–Industrial applications. I. Title. II. Series.
TA1677.B69   1987   621.36'6   87-20660

© Springer-Verlag Berlin Heidelberg 1987
Softcover reprint of the hardcover 1st edition   1987

The text was word-processed using PS^{TM} software and was printed with a Toshiba P321

2153/3150-543210

Preface

This text aims at providing a comprehensive and up to date treatment of the new and rapidly expanding field of laser processing of thin films, particularly, though by no means exclusively, of recent progress in the dielectrics area. The volume covers all the major aspects of laser processing technology in general, from the background and history to its many potential applications, and from the theory to the necessary experimental considerations. It highlights and compares the vast array of processing conditions now available with intense photon beams, as well as the properties of the films and microstructures produced. Separate chapters deal with the fundamentals of laser interactions with matter, and with experimental considerations. Detailed consideration is also given to film deposition, nucleation and growth, oxidation and annealing, as well as selective and localized etching and ablation, not only in terms of the various photon-induced processes, but also with respect to traditional as well as other competing new technologies.

This monograph came to be written upon the invitation of the publisher and is based on a variety of lectures and review papers given on the subject over the past few years. It was decided to avoid producing a volume containing mainly computerized tables and bibliographic listings devoid of the necessary description and discussion. Thus, the book is intended not only as a snapshot review but also as an introductory text for those new to the field. It is written for engineers and scientists alike, from graduate to senior professional level. The subject is approached on a pedagogical level, rather than as a heavily specialized high level review. By its nature, the laser processing field is multidisciplinary, drawing widely from materials and laser science, from electronic engineering and inorganic chemistry. Therefore an attempt has been made to reduce the amount of technological jargon and technical terms, and necessary chemical formulae are defined where they are first introduced.

Many people have contributed to the production of this book. Firstly, I would like to thank Dr. H. Lotsch of Springer-Verlag for his patient assistance and invaluable advice over many months of preparation. I am also indebted to many of my colleagues in the field of laser processing for all the valuable discussions over the past years, and for their kindness in providing original figures and references from their work. Special thanks go to those who habitually include a thorough reference list in their papers, and to those who have written review articles.... Without these, this book would have taken much longer to write. I am also grateful to Lindsey Gall for typing earlier drafts of the manuscript, to Bridget Bradley for making many last minute adjustments and updates, and also to Betty Smith for the necessary secretarial back-up.

I am also indebted to colleagues at the Heriot-Watt University and Hughes Microelectronics Ltd. in Scotland, and at the North Texas State University, for stimulating discussions and collaborations, and more recently my colleagues in the department of Electronic and Electrical Engineering at University College London for their generous support and encouragement over the past two years. Finally, I would like to thank my wife, Ann, for her help during the preparation of this monograph, and also for her support and great patience in tolerating the eccentricities and odd hours of an absentminded and somewhat preoccupied husband over the past few years.

London, July 1987 I.W. Boyd

Table of Contents

1. Introduction

This introductory chapter is aimed at providing the reader with the background as to why laser processing has attracted so much attention throughout the world in recent years. After a brief summary of the history of the field in relation to the more traditional industrial applications of lasers in cutting, drilling and welding, the special advantages afforded by laser processing are discussed. Here, the unique properties of lasers are introduced, and extended to show how special new processing conditions can be achieved. In this respect, several areas of potential application of these new processing procedures, particularly in microfabrication technology, are introduced,

1.1 Historical Background

If one examines the present market-place closely, it is clear that alongside research and development, the area of materials processing is the largest application of lasers, regularly capturing some 20% of the dollar sales in recent years. Furthermore, this field is strongly expected to continue increasing very rapidly during the remainder of this century [1.1].

The reason for this optimistic forecast is the ever increasing demand from manufacturers for production techniques that are cheaper and faster, or more precise than existing manufacturing methods. Additionally, the laser offers a most unique set of processing parameters that are not available with other existing technologies, such that new structures and new materials can be formed. Also, in terms of application on the production lines, they are completely compatible with the control technology of existing modern digital electronics.

In the field of materials processing, lasers have already found applications in welding, bonding, cutting, drilling, and shaping of a wide variety of materials, including various metals, ceramics and cloth [1.2-8]. There is now an additional and increasing interest in the many real and potential applications in areas of high technology such as micromechanics, integrated optics, semiconductor device fabrication, and various aspects of chemical processing [1.8-22].

Materials processing with lasers can be divided into two categories. Firstly, the area of non-reactive processing, where the energy from the laser beam serves to modify in some way the structure of the workpiece. Phase changes, permanent or otherwise, may be involved but basically there will be no new compounds present in the final product. This includes techniques such as cutting, marking and hardening, as well as laser annealing, where the original defective structure of a material can be reorganised into an ordered crystalline form in periods of the order of microseconds and less. The opposite phase transformation, that from the crystal to an amorphous structure, is also possible using laser radiation [1.23]. These truly remarkable laser-induced phase transitions were the subject of intense investigations worldwide in the mid-1970's and early 1980's in the area of laser interactions with semiconductors [1.8-12]. In fact, the high level of activity in recent years in this area of laser interactions with and modifications of thin films is a consequence of the publication of a series of controlled experiments in laser annealing during the 1970's by several Soviet research groups [1.24-27]. Figure 1.1, for example, shows how the initial attraction of laser annealing has decreased somewhat over the 1980's from a peak around 1981. In contrast to this trend, there has been a corresponding upsurge in interest in related areas of laser processing of the

2

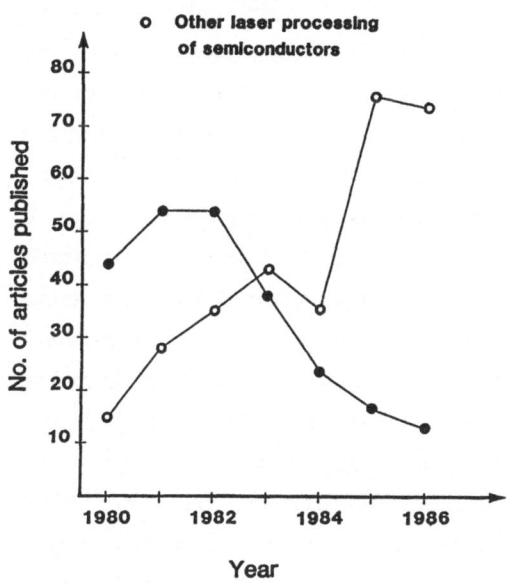

- Annealing
o Other laser processing
  of semiconductors

No. of articles published

1980 1982 1984 1986

Year

Fig.1.1. The number of papers published on a yearly basis in the Journal of Applied Physics, Applied Physics Letters, and the Journal of Vacuum Science and Technology since 1980 in the areas of laser annealing of semiconductors, and laser microfabrication of thin films for semiconductor technology [1.29]

wide variety of layers that constitute present–day semiconductor devices [1.28–36].

Since the area of non–reactive processing involves principally the action of thermal mechanisms, it is defined here as "conventional" laser processing. The treatment is usually performed in vacuum or in a non–reactive atmosphere, and the temperature distribution in the material determines the rate at which it proceeds. This distribution will, in turn, be determined somewhat by the optical absorption and the thermal diffusivity of the irradiated material, and also by the latent energies involved in any significant phase transformation, such as surface hardening, melting, evaporation, amorphization, or crystallization, etc.

When presented with the optimum environmental conditions, these conventional laser processing techniques can sometimes be applied to forming new compounds. An example of this is laser annealing of a material containing implanted or adsorbed layers, or situated within an appropriate gaseous environment. This introduces a new area of laser processing, namely, laser–assisted chemical processing (LCP).

3

LCP is, in fact, even more general. It can take place at gas/solid or liquid/solid interfaces, in adsorbed layers, or within the bulk of the material itself [1.13-21,28-36]. Consequently, material deposition, etching, doping, and compound synthesis are possible. A particular attraction of this technique is that much reduced beam irradiances can be used since surface melting is not required. The chemical reactions induced are either thermally controlled at temperatures ranging from 400°-1000°C, or photochemically regulated, where the individual photons of the laser beam selectively excite bonds and thereby initiate particular reaction pathways. As in conventional laser processing, the reactions can be confined locally, or they can be extended over some predetermined geometrical pattern. Alternatively, depending upon the size of the laser beam employed, they can cover large areas as with some traditional film preparation techniques such as chemical vapor deposition (CVD).

It is the application of laser beams in the formation of thin films and microstructures on various surfaces that will be the focus of attention of this book. Extensive investigations in this area have already produced thin layers and microstructures of metals and semiconductors, and various types of insulating films. Insulator processing has arguably attracted the most attention, and has certainly introduced by far the most varied and unique laser-assisted techniques in this area. These include alloying, annealing, polymerization, crystallization, deposition, curing, evaporation, plasma formation, oxidation, nitridation, etching, ablation and trimming, which in their various ways can be applied to either extended or patterned insulator formation. All of these processes will be covered in this book. Both the fundamental mechanisms as well as a thorough review of the work published in each area will be presented.

It is important to point out at this stage that the various well-known and well-established fields such as laser welding and laser cutting, which have already been widely adopted into the industrial environment, will not be discussed at length

4

here. The reader is referred to the literature [1.2-7] for more information on these areas. The rather heavily researched area of laser annealing will likewise not receive significant attention except where new compounds have been undeniably formed as a consequence. Again the reader is referred to several papers and books on this topic [1.9-12]. Neither is it our aim to discuss the more traditional field of photochemistry for large-volume homogeneous chemical reactions, which has been studied for many decades, if not centuries. The uses of lasers to study the fundamentals of certain reaction pathways, to isolate specific isotopes in various mixtures, or as diagnostic instruments are already well documented in [1.37-42]. The field of heterogeneous chemical processing with lasers has been active for more than a decade, with some isolated papers having been published before 1970. As will be seen, however, the vast majority of work in the field has only been produced since the early 1980's.

For the remainder of this introductory chapter, we will expand further on the background requirements demanded in this field. In particular, the reasons why the laser has been chosen as a possible tool for future processing techniques will be discussed, highlighting the unique and previously unavailable processing conditions that it now presents. Potential applications in the industrial environment, especially the microelectronics industry, will also be described here, and again more fully in Chap.8.

1.2 Advantages of Laser Technology

The desirability of these sophisticated and sometimes exotic light sources in materials processing can be understood by considering the basic properties of these instruments. Perhaps the most striking behaviour of the laser beam to the observer is its directionality and highly collimated nature, with a very low angle of divergence, typically 0.2-10 radians. This has the important consequence that large amounts of energy can be rather easily manipulated and directed onto almost any location. Furthermore, this also means that it can be collected and propagated very large distances from its source extremely efficiently without any energy being lost through divergence out of the beam path. By comparison, conventional sources of radiation such as lamps or furnaces, emit energy over the complete solid angle of 4π steradians and very elaborate collection geometries are required to minimize energy losses, which still tend to be significant.

Another desirable property of laser radiation, governed by the fundamental laws of diffraction, is the ability to focus to extremely small dimensions. In fact, a single-mode beam of laser light can, depending on the quality of the optics used, be focused down to a spot almost as small as the wavelength of the light itself. Indeed, this is the fundamental limit of focusing, which cannot be surpassed irrespective of the quality of the lenses used. Traditional light sources cannot compete in this regime. With laser processing, it is still possible to define relief on a flat surface which is smaller than the diffraction limit of the beam itself. The chemical reaction induced by the laser beam need only be a strong linear function of the incident fluence or the induced temperature profile for this to be possible, but the more strongly nonlinear this relationship is, the easier it will be to "defeat" the lower limits of resolution dictated by diffraction. Flat-topped beam profiles will also be able to create this effect. It is not difficult to see that the larger the nonlinearity involved, the smaller will be the dimensions of the features processed. These points are discussed more fully in Sect.3.2.

We can see therefore that laser beams can be rather easily directed along a well-defined path and focused down to much smaller spots than radiation from other sources can, and this indicates that in areas where the ultimate optical resolution is desirable, the use of well-defined laser beams will be most

useful. A further consequence of this extreme degree of focusa-
bility, together with the total amount of power available in
some laser beams, means that extremely high photon fluxes can
be obtained in small volumes. This holds particular promise for
more efficient chemical reactions than those previously attainable.

The advancement of digital electronic control techniques,
together with the ease of manipulation of laser beams, enables
the possibility of novel direct-writing applications. In particular, by controlling the velocity and direction of the reac-
tion-controlling laser beam, lines, or generally almost any
pattern can be defined directly on a substrate, through local-
ised deposition, growth, etching, ablation, or whatever method
is desirable. This offers for the first time a single-step
method for engraving a micrographic structure as used in micromechanics or microelectronics manufacturing. In contrast to
this, standard present-day techniques require several serial
production steps to achieve essentially the same result. Details of this technique will be discussed in more detail in
Chap.7.

The laser possesses even more useful properties, which
further increase its future potential uses in this area. The
monochromatic nature of laser light may be fully exploited by
the resonator cavity around the lasing medium which ensures a
high degree of frequency selectivity from the natural emission
frequencies of the lasing medium itself. In fact, frequency
stability better than 1 MHz is achievable. Some lasers can be
readily tuned over a considerable range of discrete wavelengths, although it must be said that the overall degree of
spectral availability is not yet close to that ideally desired
by the spectroscopists. Nevertheless, tunability and monochromatic properties present a most unique method of reaction control never encountered before the emergence of the laser.

Figure 1.2 shows a selection of the laser wavelengths commonly used for materials processing. A more complete listing is

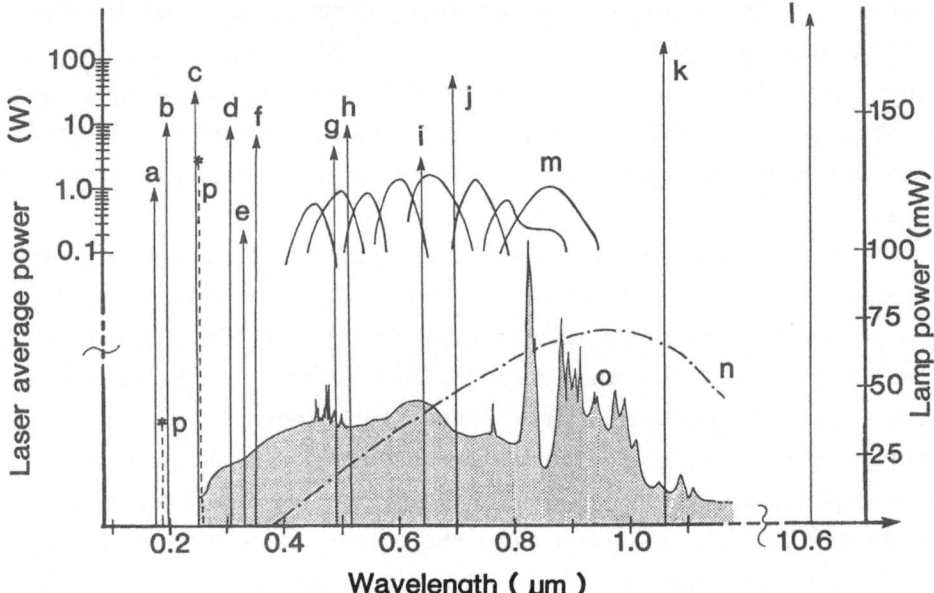

Fig.1.2. An indication of the range of wavelengths and very approximate optical powers presently available from the common commercial lasers and incoherent lamp systems; (a–d) Excimer lasers, (e) N_2 laser, (f–h) argon ion laser, (i) krypton ion laser (strongest line), (j) ruby laser, (k) Nd:YAG or Nd:glass laser, and (l) CO_2 laser. The traces shown by (m) are a series of tuning curves available from various dye laser systems pumped by some of the lasers mentioned above, and/or flash lamps. An indication of the relative power available from each system is given by reference to the logarithmic scale on the left of the diagram. Also shown is the near black body spectrum from a tungsten lamp (curve n) and the more heavily featured spectrum from a Xenon arc lamp (shaded, trace o). The discrete dotted lines (p) are the strongest emissions available from a Hg lamp. The linear scale for the lamp spectra is on the right of the diagram, and both scales are aligned at 100 mW [1.29])

given in Sect.3.3.4. A much wider range of wavelengths can be obtained by employing the UV and visible lasers to pump a large number of commercially available dyes. With this spread of frequencies available, a variety of physical and chemical mechanisms are on hand to induce many kinds of reaction phenomena for thin film processing. Also shown in Fig.1.2 are the typical spectra of hot W filament and Xe arc lamps, and the strongest lines available using a Hg discharge. The relative powers of each are also indicated, highlighting the increased energy levels that lasers now offer. These incoherent light sources have, in fact, been used since the turn of the century in the study of gas-phase decomposition, and even today are being utilized in the same fashion, as well as for possible large area film photodeposition applications.

Finally, we note here the variety of laser pulse durations that are available for laser processing. For example, we are now entering the era where laser pulses as short as 8 femtoseconds ($8 \cdot 10^{-15}$s) are being generated. Although these pulses are not readily available for photochemical processing, (and, of course, it is not obvious that they will ever be required for such applications) the generation and application of picosecond and nanosecond laser pulses allow very novel and unique territories to be investigated. We will see in this volume that nanosecond and microsecond irradiation procedures are actually quite commonly used to produce microstructures. The continuous-wave (CW) laser is more commonly applied in a great many processing systems, where irradiation times from fractions of one second to several minutes or longer are desirable. We will explore in greater detail in the next chapter, the importance of the duration of the laser pulse in determining the ultimate temperature rise induced in the solid.

1.3 Requirements for Laser Processing

We have already discussed above some of the more notable characteristics desirable for potential applications in mass production lines. One of the main considerations in this respect would be, of course, economics. The rapid and efficient processing steps that lasers can be programmed to perform have already resulted in a steady increase since the early 1970's in the number of actual laser applications on various production lines worldwide. A typical listing of some of these includes [1.2-12]:

1) Machining (scribing, cutting, drilling, shaping) of metals, semiconductors, and insulators, including brittle ceramics and synthetics.
2) Welding and alloying.
3) Surface hardening, mainly of metals.
4) Vaporization trimming of electronic components.
5) Stripping of material coating, e.g. insulation from wires.
6) Introduction of damage sites on silicon wafers to adsorb impurities and defects during high temperature processing.

The economic constraint is clearly one of the most important. For example, at present the CO_2 laser has found the great bulk of industrial applications [1.43] because of its very efficient operation in terms of energy conversion from electrical to highly coherent radiation (values around 20% are typical). What is more, these lasers are available commercially at powers in the multikilowatt range, and are without doubt the most reliable and durable systems yet available for materials processing.

Although economic viability is clearly essential for these laser applications to get off the ground and eventually displace existing and more conventional technologies, the laser also presents novel and unique processing characteristics as we have seen in the previous section that give rise to operating conditions never previously available. Perhaps the most desirable aspect of the general area of beam processing of materials is the total elimination of mechanical contact during the structural transformation induced on the sample surface. As higher levels of technology become more in demand an overall decrease in levels of contamination and material handling will become increasingly necessary, and again lasers offer such clean processing alternatives. Unlike other beam sources of energy, such as ion and electron beams, for example, laser radiation can be propagated through a great variety of intermediate media, or be chosen so as to be strongly absorbed where necessary.

There are further notable advantages of laser chemical processing which are worth mentioning. Firstly, it is clear that laser processing is not limited to the traditional planar structural configurations of many technologies. It is quite possible that three-dimensional structures may become a reality for the near future. Secondly, the choice and tunability of the laser wavelength gives the user some degree of control over the process and the actual processing depth. The penetration of the radiation can be varied across the surface of a sample, thereby giving a new degree of selectivity to the style of the delineation.

Reductions in the amount of high temperature processing are also possible with this technology. There are examples in micromechanics, e.g. in microdetonator welding where reduced temperatures, or reduced times at high temperatures are desirable and laser processing has been investigated. But it is the area of integrated circuit technology which is potentially the largest single benefactor of this particular processing condition. As the dimensions of microelectronic devices shrink towards and soon below the 1 μm barrier, it becomes increasingly important to reduce the amount of high temperature processing during the manufacture of the circuits. High-temperatures promote the diffusion of carefully prearranged dopants in the devices and also various degrees of wafer warpage, both of which are counter-productive in submicrometer processing. Since, for example, the oxide layers are extremely important in microelectronic circuitry, it would be of immense technological importance if these or similar quality insulating films could be prepared rapidly and at reduced lattice temperatures.

A great majority of the steps in present-day manufacturing microelectronic circuits require high temperature processing. By considering only the established silicon technologies, we can see that not only oxidation, but also annealing, diffusing, doping, deposition and contact formation are all performed at temperatures where it would be desirable to minimise both the times and the temperatures involved. Lasers offer a wide range of alternative possibilities for many of these operations, all of which have been intensely investigated over the past few years. Furthermore, lasers also enable the direct formation of microstructures, an alternative to forming patterns by conventional lithographic methods combined with large area processing techniques such as CVD, PCVD, or plasma etching. These techniques presently reqires up to 15 individual processing steps including spinning, baking, aligning, exposing, stripping, and etching. It is this prospect in particular that has been responsible for the unequalled enthusiasm with which this field is being investigated.

It is not only microelectronics, but also optoelectronics, the formation of fine-scale piezoelectric structures, and many other high technology areas that could benefit from the use of these techniques. One of the main reasons, for example, for the sluggish progress of GaAs-based technology are the severe problems encountered during high-temperature processing, and the poor performance of many devices incorporating such processing steps. Alternative and extremely expensive manufacturing methods are nevertheless continually being developed. Probably one of the more demanding expectations of the near future will be the incorporation of both GaAs- and Si-based technologies on the same integrated circuit, and again laser-assisted techniques could be of potential use in this area.

More specialised areas, such as custom building of the "first draft" of masks for integrated circuits, as well as discrete repair and modification of each plate, or of the processed circuit itself, can also be influenced by these laser processing methods. Such techniques combine to form another area of laser application, namely laser personalisation.

1.4 Outline

This book is set out as follows. Chapter 2 is devoted to the fundamentals behind the various laser-induced reactions. It deals primarily with the absorption and redistribution of the incident optical energy within the gas and solid phases. The interaction of laser radiation with surfaces is also considered, and photon-stimulated excitation leading to adsorption and desorption is described. In addition to the direct excitation of selective vibrational and electronic modes of the systems, the eventual degree of heating induced in solids is formulated for specific materials under particular conditions. These heating models can be applied for both pulsed and continuous laser radiation. Chapter 2 also deals with some initial work on rates of the chemical reactions induced by low and high irradiation levels. Although fundamental in nature, this chapter is aimed at providing a fairly introductory background to understanding the more important basic principles involved in LCP.

Chapter 3 reviews the various aspects of experimentation. The properties of Gaussian laser beams are introduced, and their effect on the ultimate resolution available for pyrolytic and photolytic reactions is considered. A variety of geometrical configurations is discussed together with the design criteria for the various processing modes and their limitations. The chapter ends with a summary of the range of commercially available laser systems that are commonly used for laser processing of thin films.

In Chap.4 an overview of laser induced oxide growth is presented, together with a brief summary of the present understanding of oxidation in general, and of silicon in particular. The uses of laser radiation to controllably heat up, in the solid phase, metals and semiconductors to eventually induce oxidation are described. In particular, photon enhanced oxidation will be reviewed. Chapter 5 deals with the application of lasers to anneal and melt thin surface layers in order to produce new compounds. The techniques included involve both direct implantation of the required species for the reaction as well as melting or annealing in an environment containing the necessary atoms.

Laser-induced deposition by photochemical and photothermal mechanisms is the topic of Chap.6. In addition to the oxide and nitride layer deposition, organic polymer formation is described. In Chap.7 the various aspects of selective material removal by laser irradiation are described. These not only include chemical etching by photothermal and photochemical mechanisms, but also ablation of organic layers, and cutting and drilling of metals by means of laser-induced oxidation, and also fine-scale trimming of resistive materials. The field is summarised in Chap.8, where the possibilities for future applications are discussed.

Although the materials and applications discussed in this book are unavoidably connected with silicon VLSI technology, where the world's attention is currently focused, where appropriate, all the optoelectronic materials studied have been included and referenced. Although not so heavily researched, it is possible that many of the techniques described here are perhaps more appropriately applied to such films and devices.

This book is written both as a basic introduction to, and simultaneously a snapshot review of, the field of insulator and oxide growth and patterning using lasers. Every effort has been made to reference and discuss as many publications as possible. It is hoped that this volume serves as a guide not only to students but also those in university and industry alike who are interested in the progress and development of the techniques and understanding of laser processing. With its collection of nearly 500 references, it will hopefully also serve as a useful summary reference to those already in the field.

2. Interactions and Kinetics

Most of the theoretical detail contained in this monograph is confined to this chapter. Here is discussed the interaction of electromagnetic radiation, from the ultraviolet through the visible to the near infrared, with matter. The treatment of laser excitation of gases and solids through vibrational and electronic transitions, and the subsequent energy transfer or bonding modifications, is from an elementary level. Particular attention is given to the special properties of surfaces; the mechanisms of adsorption, desorption, and nucleation, so important in this area, are discussed. A more practical review of some useful modelling work in the areas of laser induced heating and the various reaction modes and film growth rates induced by lasers is also presented.

2.1 Laser Excitation of Matter

In this chapter, the interaction of laser radiation with solids, with gaseous systems, and with adsorbed species, with relevance to thin film formation, will be discussed. Subsequent energy redistribution processes will also be described, as will the particular forms of the characteristic chemical and physical reactions that are consequently induced on an adjacent surface, or within the irradiated material.

The interaction of laser radiation with matter has been a topic of intense study over the past 25 years, and continues to attract as much interest as ever with the rapidly expanding availability of new materials and new laser beam characteristics. We are extremely fortunate in this field of research that progress on the understanding of the structure of matter has been so rapid in the past century, such that the various features of the observed optically-induced transitions can be readily defined, and reasonably well understood.

Laser radiation from the ultraviolet to the infrared interacts primarily with the electrons of an atom or atom system, and sometimes, though less commonly, with a particular set of atomic vibrations, or phonons. Furthermore, we also know that the frequencies and energies associated with this reasonably extensive band of electromagnetic radiation do not induce nuclear disturbances, or even affect the energy levels of the inner core electrons of an atom. These photons do however readily interact with the outer (bound or free) valence electrons of the atoms. Clearly then, the optical properties of any material will be mainly affected by the nature of its outermost electron(s) and consequently we can generally predict typical optical characteristics of most materials by knowing their electronic configurations and/or their present state of matter (i.e. liquid, solid, vapour etc). Moreover, the behaviour of the optically excited system after absorbing the laser radiation will also depend on the atomic arrangement, as well as the nature of any other nearby species.

It is important at this stage to introduce the concepts of pyrolytic, and photolytic reactions. The former is ideally controlled by thermally stimulated mechanisms, while the latter is initiated by direct photonic stimulation of a particular bond, after which a reaction will occur either as a result of the disruption of the bond, or simply as a consequence of the atom being in an excited state. In many of the laser-induced reactions recorded in this area of research, both thermal and nonthermal processes are present, and we therefore mostly consider the reaction to be mainly thermochemical or mainly photochemical. There are nevertheless sometimes quite obvious and discernible features that distinguish between the two mechanisms and lead to particular limitations and/or advantages of either process.

In this chapter, the basic processes of gas phase and solid phase excitation by laser radiation are described. Particular attention is given to surfaces and adsorbates in this respect. The timescales for various energy redistribution mechanisms, and eventual lattice heating are also presented while several formalisms for estimating temperature increases on the irradiated solid surfaces are summarized. Additionally, some methods for modelling the induced chemical reaction rates are described.

2.2 Laser Excitation of the Gas Species

It is well known that compared with solids, atomic and molecular species only exhibit a relatively small number of excited states that are easily accessible by optically induced transitions. The transitions allowed between rotational, vibrational, and electronic levels, within the selection rules of quantum mechanics, result in a series of discrete, and sometimes extremely narrow lines in the absorption spectrum of most gases, which are often used as an identifiable fingerprint for the species.

The approximate order-of-magnitude ratio of energies associated with electronic, and vibrational and rotational transitions are about $10^6:10^3:1$. In spectral terms we may say that vibrational changes give rise to a coarse structure, while rotational changes produce a fine structure on the more energetic electronic spectra. Although pure rotational spectra are exhibited only by molecules possessing a permanent dipole moment and the vibrational spectra require a change in dipole during the motion, electronic spectra are exhibited by all molecules since changes in the electronic distribution in a molecule are always accompanied by simultaneous changes in the dipole.

Optical excitation of an atom or molecule may lead either to a stable but reactive species, or to complete dissociation into a volatile radical. Consequently, a chemical reaction may proceed by means of film deposition or material removal. Here we will discuss separately the mechanisms of selective laser photodissociation, and then the process of vibrational-rotational excitation of the gas phase species.

2.2.1 Selective Vibrational Excitation

When atoms arrange themselves together to form a molecule they do so by means of some specific internal electronic timesharing. The nuclei of each atom attract the surrounding electrons in the complete molecule while simultaneously they repel the other nuclei, resulting in a balanced equilibrium of interatomic spacings and charge distribution. External influences may disturb this equilibrium resulting in extensions and compressions of the bonds, much like the behaviour of a spring. This rather simplistic analogy actually serves as a good initial starting point for modelling small elastic deviations from the mean bond length. Real molecules, however, do not follow exactly the rules of simple harmonic motion, and when stretched, for example, tend to eventually dissociate into their component atoms. An empirical expression known as the Morse function which approximates the general behaviour of a diatomic molecule, shows that the energy can be written as

$$E = D_{eq} \left\{ 1 - \exp[A(r_{eq} - r)] \right\}^2 \qquad (2.1)$$

where r is the internuclear spacing, A is a constant, D_{eq} is the dissociation energy, and r_{eq} is the equilibrium internuclear spacing. When this expression is substituted into the Schrödinger equation it gives rise to a set of allowed vibrational energy levels for this anharmonic oscillator, as shown in Fig.2.1. It can be seen that there is a distinct anharmonicity in the pattern. If we take the spectrum of HCl as an illustra-

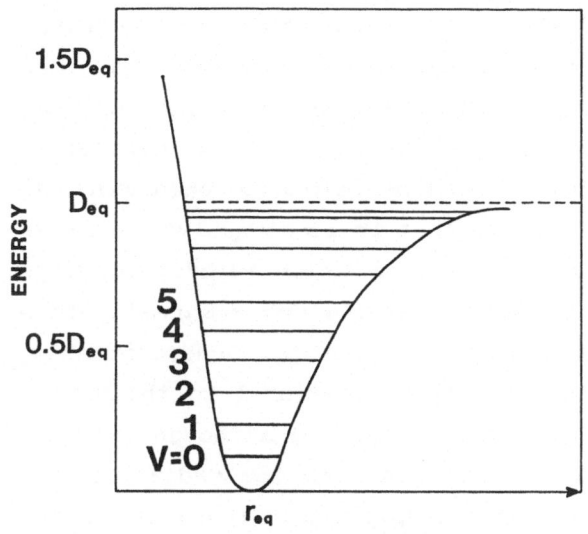

Fig.2.1. The allowed vibrational energy levels for a diatomic molecule undergoing anharmonic oscillations.

tion of the anharmonicity, and we know from measurements that there are absorption lines with decreasing intensities around 2886, 5668, and 8347 wave numbers, the anharmonicity constant, χ_e, can be calculated knowing that

$$\omega_e(1-2\chi_e) = 2886 \text{ cm}^{-1} ,$$
$$2\omega_e(1-3\chi_e) = 5668 \text{ cm}^{-1} ,$$
$$3\omega_e(1-4\chi_e) = 8347 \text{ cm}^{-1} . \tag{2.2}$$

From these equations we find that $\chi_e = 1.74 \cdot 10^{-2}$, and $\omega_e = 2990 \text{ cm}^{-1}$. Note that because of this quite appreciable anharmonicity in the vibrational spectrum of the diatomic molecule, multiphoton absorption is not very probable in the presence of highly monochromatic light from a laser. A white light source or even a multiline laser output may be more favourable for this reaction.

The vibrational states of polyatomic molecules, on the other hand, are quite different. A nonlinear molecule can for example have 3N-6 different internal vibrations. (Linear polyatomic species have 3N-5). Thus for the diatomic molecules described above then, there can only be one fundamental vibration, which as we have seen is complicated by overtone vibrations governed by anharmonicity. Each of the internal vibrations of the polyatomic molecules is also affected by these phenomena, and furthermore, there are additional effects arising from combinations of vibrations as well as overtones, and the intensities of these can be considerably enhanced by resonance phenomena. An important consequence of this is that there now exist many discrete energy levels at lower energies which rapidly converge into a quasi-continuum at higher energies and eventually to a true continuum above the dissociation energy. Understandably, the complexity of the situation is proportional to the number of atoms comprising the molecule. Therefore although only one or two modes of vibration actually absorb the radiation, the energy so absorbed can be coupled rapidly into the quasicontinuum thereby heating up the molecule towards a state of dissociation, via a strictly multiphoton process.

At present there are no clear cases of dielectric film formation being initiated by this collisionless mechanism. However, there is strong evidence [2.1] that silicon films can be deposited in this manner, by the vibrational excitation of SiH_4 with the 10P20 line of a CO_2 laser. In the presence of some oxygen-bearing gaseous species, films of SiO_2 could most probably be formed.

Another mode by which energy is absorbed into vibrational states of a polyatomic or indeed diatomic molecule is by collisional transfer. This process could clearly compete with that described immediately above, making it quite difficult to distinguish between the two processes, or indeed to purposely select the former mode of operation. However, collisional transfer can introduce interesting alternatives to this photon assisted chemical processing. It is now possible to absorb the necessary energy via some resonance process between the laser radiation and some intermediate species, such that after some collisional scattering time, the energy is eventually transferred to the desired molecule which subsequently undergoes the necessary chemical adjustments for reaction. This technique, known as sensitization, is quite common in the field of photochemistry.

Vibrational excitation from one state to the next highest, or to another not too far removed from the ground state has been found, without actually inducing dissociation, to enhance the sticking coefficients of particular molecules. Also, when N_2O molecules are excited from the (001) to the (100) vibrational level by photons at 10.8 μm it is found that their reaction with Cu (most likely to form N_2 and CuO) could be enhanced by a factor of 5000 [2.2].

For a large number of chemical reactions, however, the activation energies are much higher than a single vibrational excitation, and in order to initiate these it is often necessary to incur multiphoton excitation, leading either to a highly excited reactive species, or to complete rupture of the bonding and dissociation of the molecule. In the next section we will discuss dissociation associated with electronic transitions, but for the remainder of this section, examples of multiphoton-induced atomic separation will be described.

The phenomenon of multiple (sometimes (n≃30) photon dissociation of polyatomic molecules in their ground electronic state has been intensely studied over recent years. For example, CHUANG [2.3] has studied the decomposition of SF_6 using a high power CO_2 laser. It was found that the F atoms formed by the photodissociation of SF_6 into SF_5 + F and the subsequent dissociation of the unstable SF_5 into SF_4 + F could diffuse in their reactive state over appreciable distances to interact with a nearby Si surface. Similarly, a CO_2 laser has been used to generate CF_3 radicals from CF_3Br to react with SiO_2 [2.4]. As pointed out by CHUANG [2.5], it is expected that this laser-induced process may be more widely used in the future for etching studies since many halogen-containing polyatomic molecules such as N_2F_4, CF_3X, CF_2X_2, where X=Cl, Br, I, etc., can be readily dissociated to produce reactive radicals for possible reactions with, for example, SiO_2.

As we have seen already the energy difference between neighbouring vibrational states in a typical molecular species is of the order of 1000 times smaller than the energy difference between adjacent electronic states, while the energy difference between neighbouring rotational states is 1000 times smaller still. For example the average line separation corresponding to transitions between neighbouring rotational states in the CO molecule is approximately 3.83 cm^{-1}, corresponding to photon energies in the millimetre (microwave) region of the spectrum. Consequently, the use of lasers to promote transitions within the rotational system of excitations is not seriously considered, and therefore such effects will not be discussed here.

On the other hand, the quantized energy of the individual photons used to induce vibrational excitations is easily available with lasers operating in the near infrared portion of the spectrum (1–20 μm). Usually in cases where only one vibrational transition is required to initiate the chemical reaction, relatively low power densities are necessary. In contrast, where multi-photon processes are desirable one usually finds that much higher beam intensities are required. Clearly, the reactions will also be affected by the gas mixture and its pressure, as well as the absorption cross-section of the photon-induced process necessary for the reaction.

2.2.2 Selective Electronic Excitation

In Fig.2.2 an example of single photon electronic excitation is shown where the internuclear distances of the upper and lower excited states are essentially equal. Quantum theory predicts this to be the most probable and strongest transition. Other transitions to neighbouring vibrational states will occur with much reduced probability.

There are situations where the internuclear distances of two neighbouring electronic states are different. Consider the case where the upper electronic state gives rise to a larger separa-

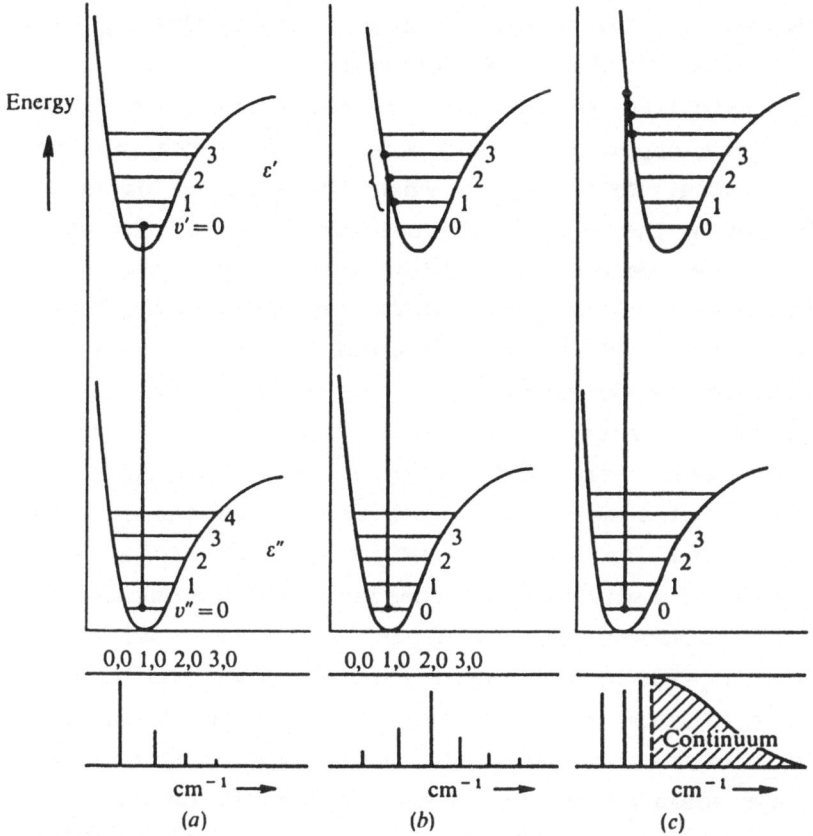

Fig.2.2. The allowed electronic transitions and associated spectra for (a) internuclear distances equal in upper and lower states, (b) upper state internuclear distance slightly greater in the upper state, and (c) upper state internuclear distance appreciably greater [2.100]. (Figure reproduced from [2.100] with permission)

tion of the atoms, i.e. the upper Morse curve in the diagram is translated to the right (Fig.2.2). A vertical transition from the ground state will now effect one of the higher vibrational levels in the ladder of the upper electronic branch. Indeed if the upper state separation is sufficiently large the most probable transition would be to some vibrational state so far up into the upper level that the excited molecule has energy in excess of its own dissociation energy. Transitions to lower vibrational states will similarly take place, but with reduced

23

probability, as will transitions to higher levels although this time these levels will form the continuum.

Another possibility for dissociation by electronic excitation occurs when the upper electronic state, unlike the previous case, is unstable. This happens when there is no definable energy minimum in the upper electronic state. Here, the whole spectrum for the system will exhibit a continuum, the lower limit of which will be the energy difference between the lowest vibrational level of the lower electronic state and that just needed to dissociate the molecule.

Finally, it is sometimes possible that the usual stable excited states are intersected by another set of unstable continuous states, at some energy level well below the expected dissociation value. In this case (Fig.2.2), there is a finite probability that a radiationless transfer can occur between the curves and "predissociation" can occur. One of the most widely studied reactions using electronic excitation and dissociation of molecules to form etch-patterned microstructures, involves SiO_2 and Cl_2. In fact, single photon photofragmentation is central to laser assisted etching of many materials involving not only Cl_2 but also Br_2 and I_2. Since SiO_2 does not absorb the argon laser radiation used in these experiments the dominant effect of the radiation is to excite the Cl_2 molecules into their first dissociative state [2.6,7].

Photolysis of $Al_2(CH_3)_6$ by UV radiation is also possible by laser-induced dissociative electronic transitions, and in the presence of N_2O or NO_2 gas the deposition of Al_2O_3 has been achieved [2.8]. Figure 2.3 shows the absorption spectra of both N_2O and $Al_2(CH_3)_6$. Both can absorb and be readily dissociated by photons at 193 nm, and can subsequently participate in a number of reactions. This reaction can also be affected by directing the UV radiation on to the surface and exciting particular surface phase species, and consequently the complicated chemistry behind the growth of these Al_2O_3 films has yet to be formulated. The effect of laser radiation on various surfaces will

24

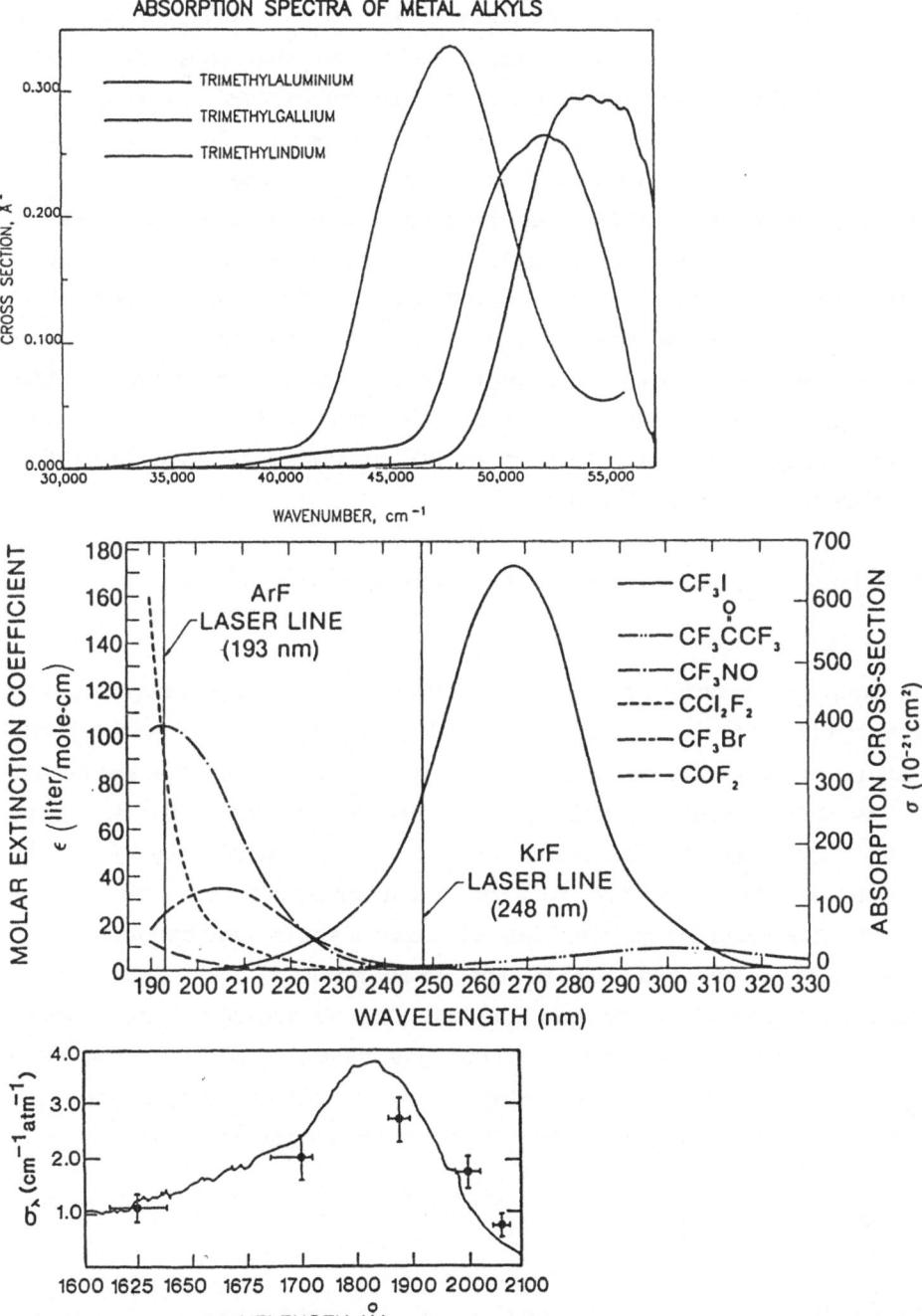

Fig.2.3. The absorption spectra of (a) some metal alkyls [2.81] (courtesy R.M. Osgood, Columbia University), (b) various radical precursor compounds [2.82], and (c) N_2O [2.83,84]

be discussed later in Sect.2.4. Figure 2.3 also shows the absorption spectra of a range of halogen-containing precursors studied by LOPER and TABAT [2.9] for laser induced etching.

In general, since these molecules are known to exhibit dissociative continua around the UV region of the spectrum, the use of frequency doubled wavelengths from the argon ion laser, or the fundamental wavelengths from the wide variety of excimer lasers is most desirable for this process. It is somewhat surprising to note the apparent lack of studies in this area using the relatively efficient frequency quadrupling of the 1.064 μm radiation from the Nd:YAG laser (to 266 nm), especially since pulse durations less than 20 ps would be available for time-resolved investigations.

2.3 Interaction of Laser Radiation with Solids

2.3.1 Metals

As mentioned briefly in the introduction to this chapter, the absorption of radiation in solids is chiefly governed by the nature of the outermost electrons of the atoms in the material. As is well known, metals are characterized by their loosely-bound outermost electrons, which are essentially free to travel around from atom to atom. As a consequence of this behaviour, the optical properties of most metals can be quite successfully described within the framework of the highly successful "free electron gas" model. Calculations of the dielectric response of such an electron gas reveal a plasma frequency ω_p at which an undamped plasma of electrons can resonantly oscillate. The frequency for which this occurs is given by

$$\omega_p = (e/2\pi)\sqrt{N_e/m_e\epsilon_0} \qquad\qquad (2.3)$$

where N_e and m_e are the number density and mass of the electrons, of charge e, and ϵ_0 is the dielectric constant $8.85 \cdot 10^{-12}$ F/m.

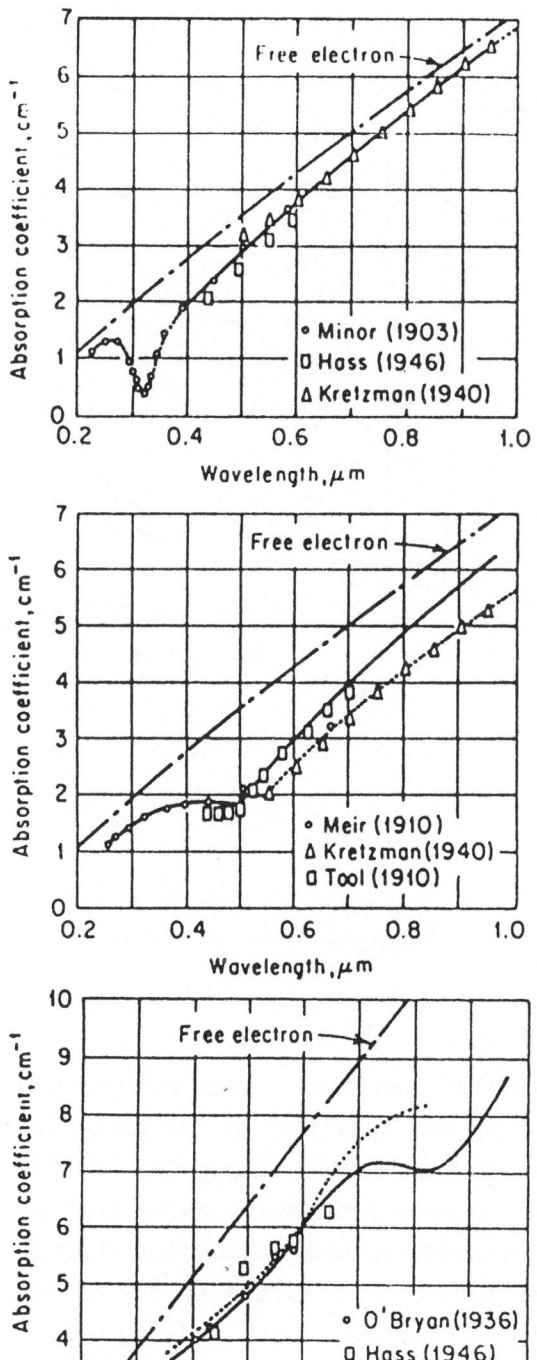

Fig.2.4. Optical absorption curves for various metals, collected from various sources [2.85, 86]. Figure to be continued on next page

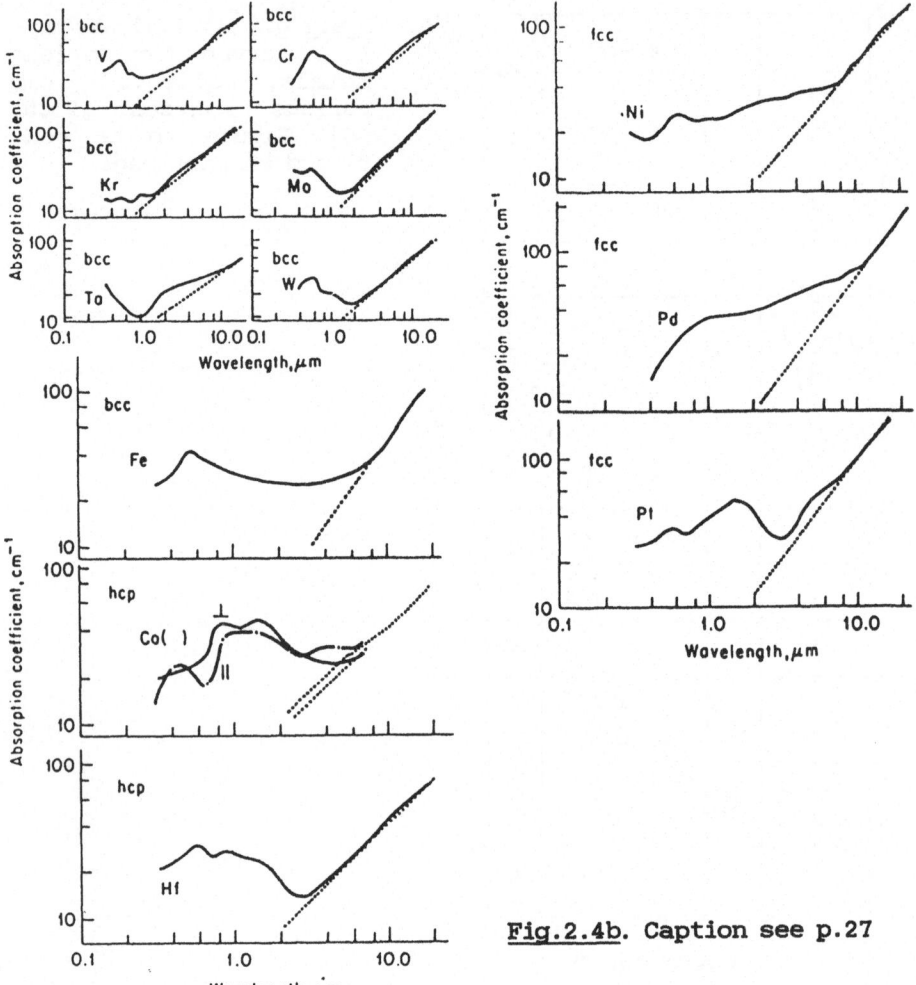

Fig.2.4b. Caption see p.27

At low frequencies, generally much lower than this plasma edge, absorption is dominated by the behaviour of the free-carriers in the material. Figure 2.4 shows the absorption coefficient of some metals. It can be seen from every curve that at wavelengths slightly beyond one micron, and longer, the absorption characteristics of the metals are virtually identical, being totally controlled by the free electrons. This behaviour can be fitted by a line of slope 2 on these plots, agreeing satisfactorily with the theory, which will be outlined later in

this chapter, that the absorption coefficient is inversely proportional to the square root of the frequency of the radiation.
Extrapolation of this relationship to lower frequencies will
bring about extremely high absorption in these metals resulting
in an absorption depth (the inverse of the absorption coefficient) so small that it is commonly known as the "skin depth".

Electromagnetic radiation whose frequency is also lower
than the plasma frequency, is however, strongly reflected from
the system, owing to the collective plasma properties of the
free electrons (Fig. 2.5). The consequence of this is that in

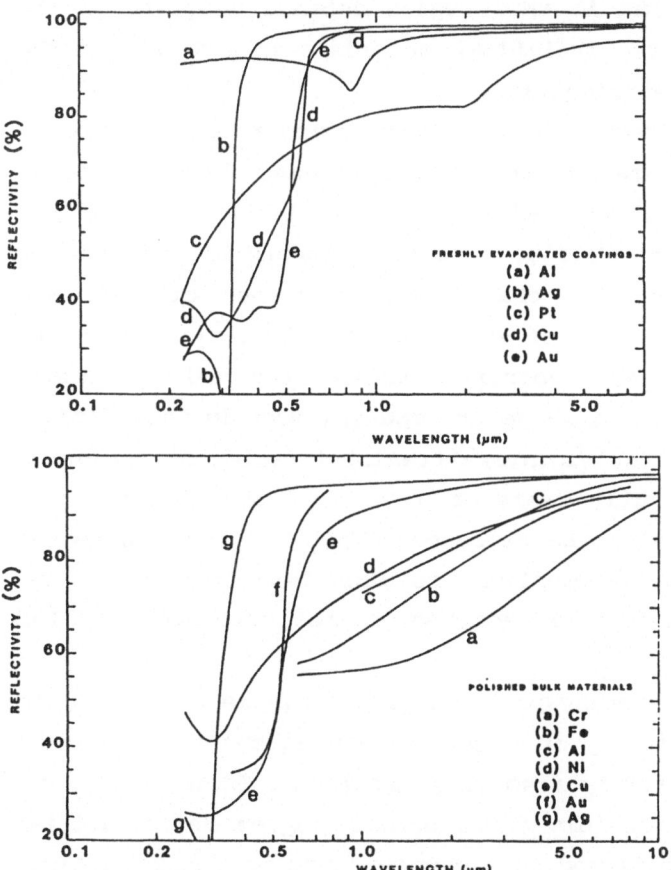

Fig.2.5. Optical reflectivity of selected metals is shown here
as a function of wavelength, for (a) freshly evaporated thin
films, and (b) polished bulk materials. (Data collected from
[2.70,85,86])

order to make use of the efficient absorption processes, extremely large laser beam intensities must be used to compensate for the large reflectivity losses. Fortunately, nature has provided us with the most efficient CO_2 laser which provides very intense radiation at 10 μm (Chap.3).

As the wavelength of the incident radiation decreases, the absorption and the reflectivity both decrease. Unfortunately, there are not usually any accurate analytical approximations that can be made for the optical properties of the metal in this region. Therefore the precise location of the reflectivity and absorption minima in this region depends uniquely on the band structure of the individual materials in addition to the plasma resonance mentioned above.

At even higher frequencies, the free carrier contribution to the optical properties becomes extremely small, and the metal exhibits characteristics related to the lattice properties. Eventually for most metals, interband absorption will occur beyond the minimum. The characteristic colour of gold and copper are a consequence of such an absorption minimum in the visible region of the spectrum, while the alkali metals (Na,Li,K, etc.) exhibit a large transparent zone in the ultraviolet. Therefore in the visible/ultraviolet region of the spectrum, one must be fully aware of both the absorption and reflection of the metal to be processed. Whilst in most cases the reflectivity and the absorption are much less than in the infrared they are much more sensitive to the wavelength of the incident radiation.

Not only is the precise position of the plasma edge vital to the optical reflectivity of a metal surface, but the condition of the surface itself is also very important. It is, for example, well known that thin films exhibit higher reflectivities than their highly polished bulk counterparts do (Fig.2.5). Also, the presence of impurities, a surface oxide, or a particular surface finish, dramatically alters the optical properties of the sample. Metals are often coated with a known thickness of

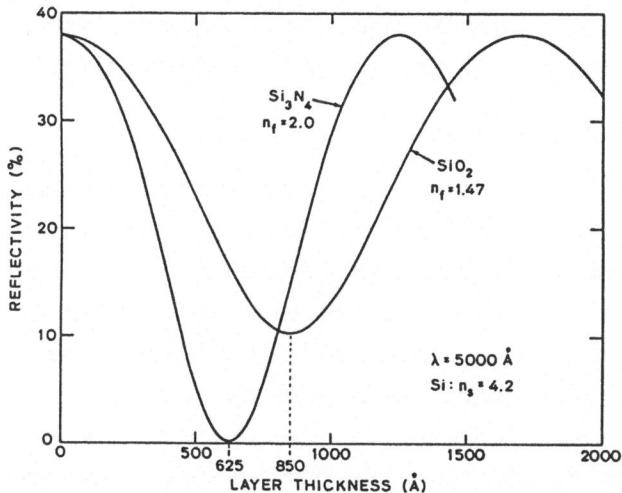

Fig.2.6. Reflectivity at 500 nm of c–Si coated with either SiO$_2$ or Si$_3$N$_4$, as a function of film thickness (From Lietoila et al. in [2.87])

some dielectric material in order to enhance the optical coupling of the energy into the bulk. This technique can either rely upon the strong optical absorption of the coating, or a multiple interference effect of the incident light through a transparent film, or a combination of both. This is shown in Fig.2.6 for thin coatings of SiO$_2$ and Si$_3$N$_4$ on c–Si (crystalline Si). Notice that the reflectivity can be reduced from 0.38 to nearly zero for a film thickness of 62.5 nm of Si$_3$N$_4$.

Once the energy has been decoupled from the laser beam into the electronic system of the metal it is rapidly redistributed amongst the other electrons via collisions and eventually to the lattice by various scattering mechanisms. Typically, times of the order of picoseconds or less are required to convert the photonic energy into random atomic motions. Thus the metal can be rapidly heated up during the irradiation event. In this way the rate of supply of energy (proportional to the beam flux and duration of the exposure) is balanced by heat losses due to thermal diffusion into the bulk, radiation losses from the surface, or convective losses. The actual temperature attained in this way can be determined using the usual heat diffusion equation, and this calculation will be presented later in Sect.2.5.

2.3.2 Insulators and Semiconductors

Unlike metals, insulators and semiconductors do not contain appreciable numbers of free carriers in their conduction bands. In contrast to having a loosely bonded electron associated with each atom, these materials can be characterized by having their outermost electrons completely filling the highest energy (valence) band. Such tight bonding is not conducive to electronic conduction. Each material, however, can be characterized by a parameter E_g which defines the amount of energy required by an electron in the highest valence state, to cross the forbidden zone (energy gap) up into the lowest level in the conduction band. For values in the region of 9 eV (i.e., well into the ultraviolet) the material is a very strong insulator. On the other hand, if E_g is closer to 0 eV, then the material may be defined as a narrow-bandgap semiconductor. The crossover point between wide-bandgap semiconductor and insulator is not well defined, but a value around 5 eV would generally be considered to be about the lower limit for insulators.

Above and below each of their bandgap energies, both insulators and semiconductors behave quite similarly, in terms of their optical properties. For radiation whose photon energy is greater than E_g, i.e. $E_p = h\nu > E_g$, electronic transitions between the valence and conduction bands are induced. This occurs in the near infrared for silicon (1.12 eV), germanium (0.67 eV), and GaAs (1.42 eV), the infrared for indium antimonide (0.18 eV), the green (2.26 eV) for gallium phosphide, just into the ultraviolet (3.5 eV) for the chalcopyrite $CuAlS_2$ and the vacuum ultraviolet (8 eV) for silicon dioxide. From one side to the other of this band edge the absorption coefficient increases from typically less than 1 cm^{-1} to well over 10^6 cm^{-1}, i.e. approaching that usually exhibited by metals in the infrared. It is also in the strongly absorbing region of the spectrum that the materials exhibit their highest reflectivity because of the inherently large refractive indices in this regime. Indeed, overall, both semiconductors and insulators strongly resemble metals at wavelengths above their individual band edges. At energies below the bandgap, the optical properties of these materials are determined by the much weaker intraband electronic transitions and by excitation of specific vibrational modes within the lattice itself. The frequency and precise location as well as the overall strength of these modes is strongly dependent upon the exact nature of the bonding arrangements, and the type of atomic species involved. This regime will be discussed in more detail below.

In order to give an indication of the complexity of accurately determining the exact optical response of any of these materials, the band structure of several of the more commonly studied semiconductors has been sketched in Fig.2.7. Also shown in Table 2.1 is a compilation of semiconductor data from various sources highlighting the subtle, yet sometimes significant, differences between the semiconductors. Taking Si as an example, the band structure shown in Fig.2.7 gives rise to a wavelength dependent optical absorption and reflectivity as shown in Fig.2.8.

Looking at this plot in detail, it can be seen that the onset of the strong absorption occurs around at 1.1 eV, which corresponds to the energy required to make an indirect transition from the top of the valence band to the bottom of the lowest point in the conduction band. Since these points are displaced in momentum space, this transition is defined as indirect, as it requires the participation of phonons to balance the difference in momentum between the carriers that occupy these states. Since the presence of phonons is necessary for the transition we would expect a strong temperature dependence of this process. Clearly the temperature dependence of the precise value of the energy gap is also important in determining the strength of optical transitions in this regime. As the photon energy is

Table 2.1 Properties of some semiconductor materials at 300 K. Many values quoted here are only approximate, particularly for the II-VI and IV-VI compounds. The data here has been taken from compilations by STREETMAN [2.90] and SZE [2.91]. (* means the material vapourizes)

	Eg [eV]	Direct/ Indirect	Lattice Constant [A]	Dielectric Constant ϵ	Density [g/cm³]	Melting Point [C]
Si	1.12	i	5.43	11.8	2.33	1413
Ge	0.67	i	5.66	16	5.32	936
C(diamond)	5.47	i	3.56	5.7	3.53	3550
SiC(α)	3.00	i	a=3.09 c=15.1	10.2	3.21	2830
Se	1.74	d	-	8.5	4.81	217
Te	0.32	d	-	-	6.25	450
AlP	2.45	i	5.46	9.8	2.40	2000
AlAs	2.16	i	5.66	12	3.60	1740
AlSb	1.6	i	6.14	14.4	4.26	1080
BP	2	i	4.54	6.9	2.97	-
GaN	3.36	d	a=3.19 c=5.19	12.2	-	-
GaP	2.26	i	5.45	11.1	4.13	1467
GaAs	1.43	d	5.65	13.2	5.31	1238
GaSb	0.72	d	6.09	15.7	5.61	712
InP	1.3	d	5.87	12.4	4.79	1070
InAs	0.36	d	6.06	14.6	5.67	943
InSb	0.18	d	6.48	17.7	5.78	525
ZnO	3.35	d	4.58	8.0	5.63	1975
ZnS(α)	3.8	d	a=3.81 c=6.26	8.6	3.98	1185*
(β)	3.6	d	5.409	8.6	4.09	1020*
ZnSe	2.7	d	5.671	9.2	5.65	1100*
ZnTe	2.25	d	6.101	10.4	5.51	1238*
CdS	2.42	d	4.137	8.9	4.82	1475
CdSe	1.73	d	4.30	10.3	5.81	1258
CdTe	1.51	d	6.42	10.5	6.20	1098
PbS	0.41	i	5.936	161	7.6	1114
PbSe	0.27	i	6.147	280	8.73	1081
PbTe	0.29	i	6.452	360	8.16	925
SnTe	0.18	d	6.328	-	6.452	-

increased, transitions can occur from deeper within each of the bands, where the density of states is significantly higher, and therefore the absorption becomes much stronger. At some point around 360 nm (3.4 eV), direct transitions, which do not require phonon assistance, may occur.

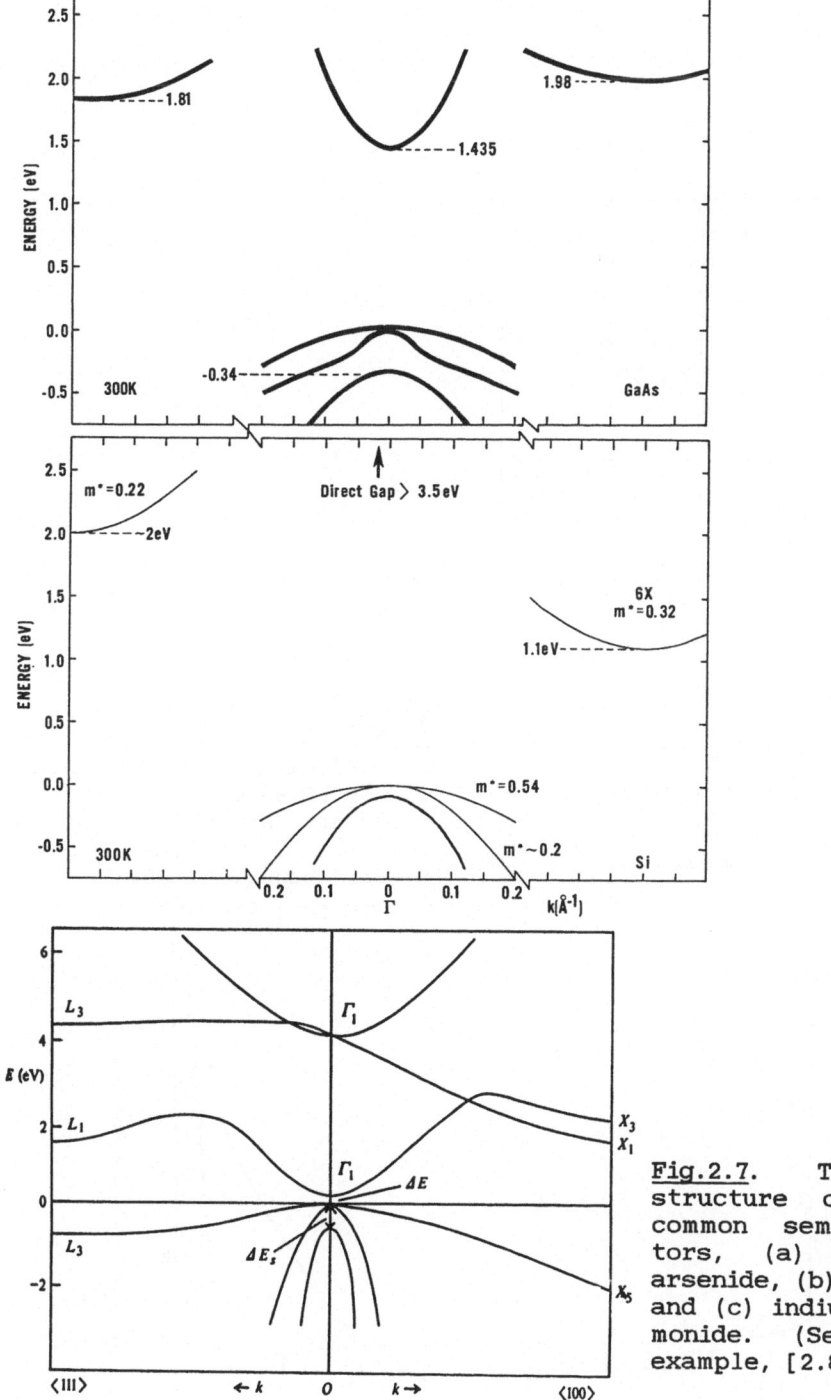

Fig.2.7. The band structure of some common semiconductors, (a) gallium arsenide, (b) silicon and (c) indium antimonide. (See, for example, [2.88,99])

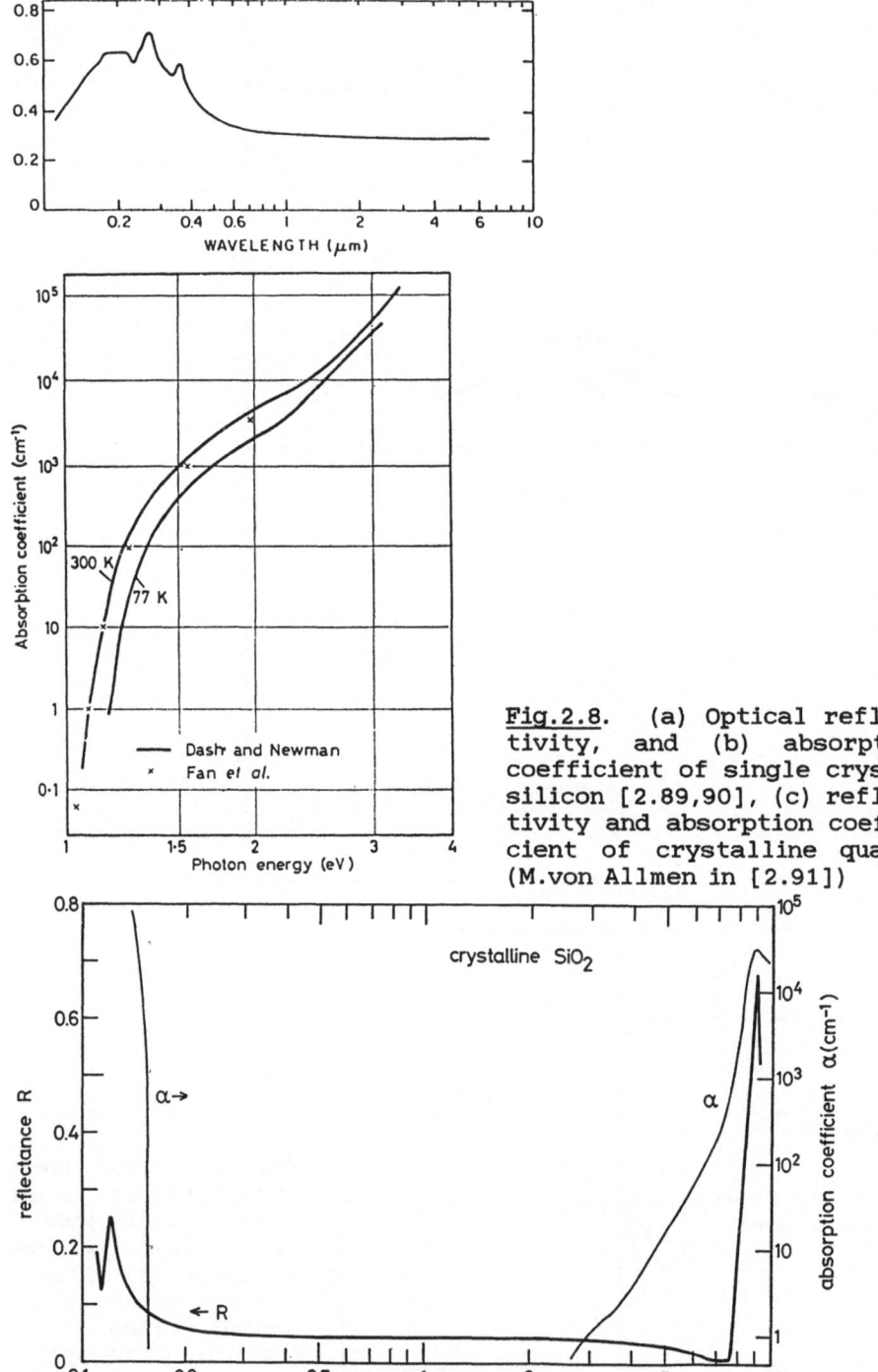

Fig.2.8. (a) Optical reflectivity, and (b) absorption coefficient of single crystal silicon [2.89,90], (c) reflectivity and absorption coefficient of crystalline quartz (M.von Allmen in [2.91])

Figure 2.8 also shows the optical properties of SiO_2, the common oxide of silicon, which can be found as crystalline quartz, $c\text{-}SiO_2$, or as a glass, $a\text{-}SiO_2$, sometimes known as silica. The bandgap associated with SiO_2 is close to 8 eV, and it is therefore an excellent insulator, exhibiting resistivities around 10^{16} Ωcm at 300 K, and dielectric breakdown approaching 10 MV/cm. These very desirable electrical properties are one of the main reasons why silicon microelectronic technology has become so successful in recent decades. More details on SiO_2 are given in Chaps.4 and 6, where laser induced growth and deposition of SiO_2 are discussed.

It is useful at this point to remind ourselves that since the radiation wavelengths of interest in this book are large compared with the interatomic dimensions, the optical properties of the solid can be described in terms of macroscopic optical constants, the refractive index n, and the extinction coefficient (sometimes known as the absorption index) k with $k = \alpha\lambda/4\pi$, where α is the absorption coefficient of the material at the wavelength λ. These quantities n and k are, of course, not strictly constant, but vary somewhat according to, for example, the frequency of the interacting radiation, and the sample temperature. Alternatively, because of the successes of Maxwell's equations, the optical properties of a material can be related to the complex dielectric constant $\epsilon = \epsilon_1 + i\epsilon_2$ and the conductivity σ of the medium. Thus

$$n^2 - k^2 = \epsilon_1/\epsilon_0 \qquad (2.4)$$
$$2nk = \epsilon_2/\epsilon_0 = \sigma/\omega\epsilon_0 \qquad (2.5)$$

where $\epsilon_1 = D/E$ and $\sigma = J/E$, with ϵ_0 the permittivity of free space, D the electric displacement, E the electric field and J the current density. Therefore, a knowledge of ϵ_1, and ϵ_2 (or σ) can lead to values for n and k, and indeed the intensity reflection coefficient R at normal incidence, given by

$$R = \frac{(n-1)^2 + k^2}{(n+1)^2 + k^2} \qquad (2.6)$$

37

JELLISON and MODINE [2.10,11] have recently measured the temperature dependence of the absorption and reflection of crystalline silicon as a function of wavelength. Their results
are reproduced here in Fig.2.9. It can be seen that the absorption coefficient increases with temperature for all wavelengths below an energy of about 3.4 eV (corresponding to the
direct bandgap of the material). More accurately, it was found
that the absorption coefficient increased exponentially with
temperature from room temperature to 700°C for photon energies
between 1.7 and 3.0 eV. This relationship can be described by

$$\alpha(\lambda,T) = \alpha_0(\lambda) \, \exp(T/T_0) \tag{2.7}$$

for any wavelength. Values of α_0 and T_0 used in this empirical
relationship have been tabulated for various commonly used
laser wavelengths in Table 2.2. The exponential relationship is
a strong indication that the absorption in this regime is mainly
influenced by the increased phonon population at higher temperatures. At wavelengths towards the shorter end of the range,

Table 2.2. The optical functions of c-Si (n and R, ϵ_1 and ϵ_2)
together with the optical absorption coefficient α, and the
calculated normal-incidence reflectivity R at several wavelengths. Also shown are the parameters relevant to the empirical fit to $\alpha(T)$ [2.10,11]

Laser	n	k	ϵ_1	ϵ_2	α [1/cm]	R
Ruby	3.763	0.013	14.16	0.10	2.4×10^3	0.336
HeNe (633nm)	3.866	0.018	14.95	0.14	3.6×10^3	0.347
Nd:YAG (530nm)	4.153	0.038	17.24	0.32	9.0×10^3	0.374
Argon (514nm)	4.241	0.046	17.98	0.39	1.12×10^4	0.382
Argon (488nm)	4.356	0.064	18.97	0.56	1.56×10^4	0.392
N2-pumped dye (485nm)	4.375	0.066	19.14	0.58	1.71×10^4	0.394
Argon (458nm)	4.633	0.096	21.45	0.89	2.64×10^4	0.416
N2-pumped dye (405nm)	5.493	0.290	30.08	3.19	9.01×10^4	0.479
Nd:YAG (355nm)	5.683	3.027	23.13	34.41	1.07×10^6	0.575
N2	5.185	3.039	17.65	31.51	1.12×10^6	0.560
XeCl	4.945	3.616	11.37	35.76	1.48×10^6	0.587

the simple equation defined above tends to be less accurate in describing the behaviour of the absorption with increasing temperature. This is considered to be most likely a consequence of a slight decrease in the direct bandgap at elevated temperatures, and the gradual domination of direct optically induced transitions over the indirect excitations. Returning to Fig 2.9a, it is seen that the absorption generally continues to increase at higher photon energies, peaking near 4.25 eV. JELLISON [2.11] summarises the various possibilities for this transition. Notice that the position of the peak shifts towards lower energies as the temperature increases, and that it also decreases in magnitude.

As mentioned above, the temperature dependence of the normal incidence reflectivity has similarly been determined over a wide range of wavelengths by JELLISON and MODINE from measurements of the optical function of silicon in the temperature range 10-972 K [2.10]. This is shown in Fig.2.9b. The peak near 3.4 eV observed in the 10 K data, shifts to lower energies with increasing temperature, and eventually disappears around 972 K, where the reflectivity increases monotonically with wavelength, from 2-4 eV. In this region, but more specifically

Table 2.2. (Cont'd)

Laser	n	k	ε_1	α_0 [1/cm]	T_0 [K]
Ruby	3.763	0.013	14.16	$1.34 \pm 0.29 \times 10^3$	427 ± 82
HeNe (633 nm)	3.866	0.018	14.95	$2.08 \pm 0.32 \times 10^3$	447 ± 62
Nd:YAG (530 nm)	4.153	0.038	17.24	$5.02 \pm 0.49 \times 10^3$	430 ± 39
Argon (514 nm)	4.241	0.046	17.98	$6.28 \pm 0.55 \times 10^3$	433 ± 39
Argon (488 nm)	4.356	0.064	18.97	$9.07 \pm 0.66 \times 10^3$	438 ± 33
N_2-pumped dye (485 nm)	4.375	0.066	19.14	$9.31 \pm 0.67 \times 10^3$	434 ± 31
Argon (458 nm)	4.633	0.096	21.45	$1.45 \pm 0.08 \times 10^4$	429 ± 34
N_2-pumped dye (405 nm)	5.493	0.290	30.08	$5.51 \pm 0.15 \times 10^4$	420 ± 69
Nd:YAG (355 nm)	5.683	3.027	23.13	$1.09 \pm 0.01 \times 10^6$	-8700 ± 5300
N_2 (337 nm)	5.185	3.039	17.65	$1.13 \pm 0.01 \times 10^6$	25000 ± 25000
XeCl (308 nm)	4.945	3.616	11.37	$1.43 \pm 0.01 \times 10^6$	4700 ± 1300

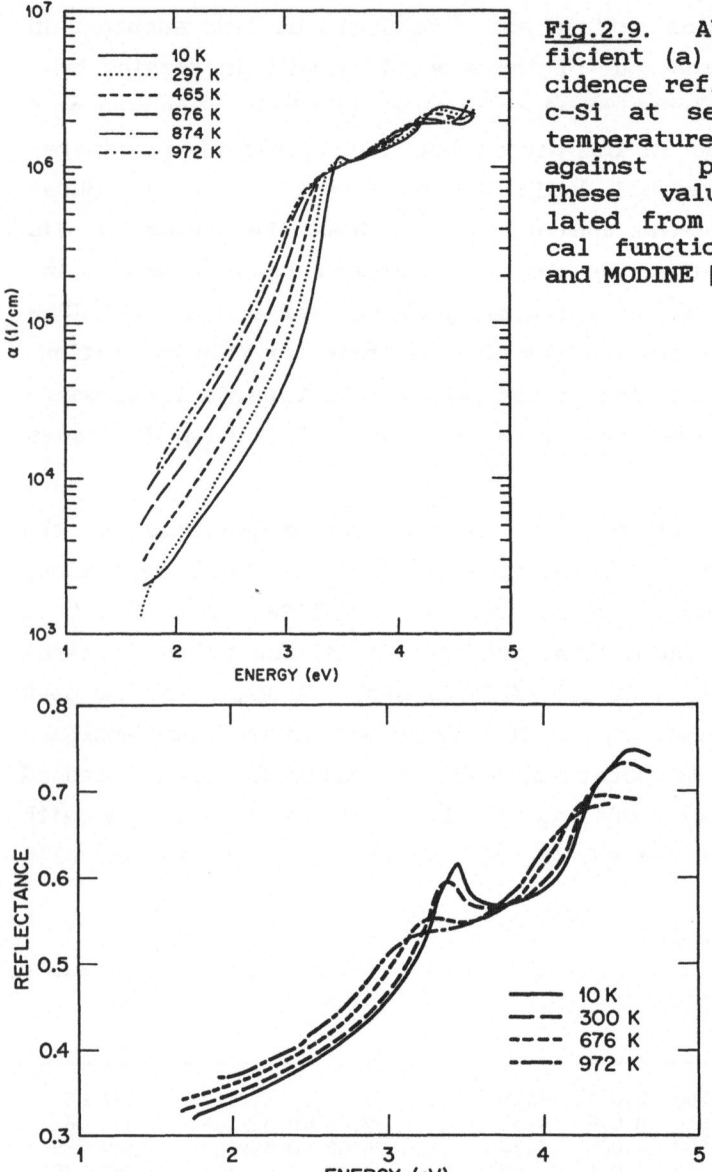

below 3 eV, R can be empirically written as

$$R(\lambda,T) = R_0(\lambda,T{=}300\ K) + 5{\cdot}10^{-5}(T{-}300) \qquad (2.8)$$

where T is in units of Kelvin. R is not a strong function of temperature in this, or indeed most other wavelength ranges.

As the energy corresponding to the fundamental absorption edge is approached, the absorption decreases rapidly, and often some structure can be detected in these regimes of lower absorption, arising from exciton effects. Absorption by excitons is very pronounced in insulating materials, particularly ionic crystals, even more so than in semiconductors, and it can lead to surprisingly strong and well-defined absorption lines similar to those usually associated with atomic spectra. There are two convenient models for describing excitons, namely the Frenkel and the Wannier models. The Frenkel exciton is pictured as an electron bound in a relatively deep impurity state that can be mobile. The electron describes an orbit of atomic dimensions, and by moving, the empty valence state, although instantaneously associated with a particular atom, constitutes a mobile hole. This model is most appropriate for solidified rare gases and some ionic crystals. The Wannier model is more applicable to the group IV semiconductors and many crystals that are partially ionic, partially covalent e.g. the III-V compounds. In this model, the electron-hole pairs are still bound by their mutual Coulomb attraction but in this case they may be separated by many atomic spacings.

Excitons exhibit discrete energy levels similar in type to those observed in hydrogen. One of the better-known exciton spectra is that of Cu_2O shown in Fig 2.10 which exhibits excitation levels up to n=5, at temperatures around 77 K. At lower temperatures, lines corresponding to n=11 have been seen.

The alkali halides are renowned for exhibiting some of the strongest exciton absorption features, sometimes with peak absorption coefficients as high as 10^6 cm^{-1}. Figure 2.11 shows the spectrum of KBr at 77 K, near to the absorption edge. It is often quite impossible to determine from such spectra where the fundamental absorption edge begins, and this must frequently be obtained from photoconductivity data. In this case the interband absorption edge was determined to be at 7.8 eV.

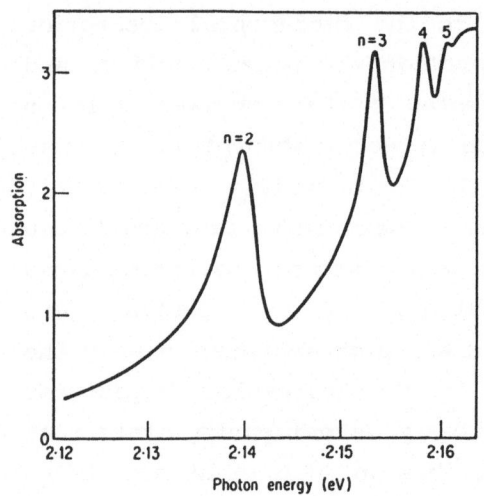

Fig.2.10. An absorption spectrum of Cu_2O at 77 K, showing the well defined exciton absorption features [2.92,97]

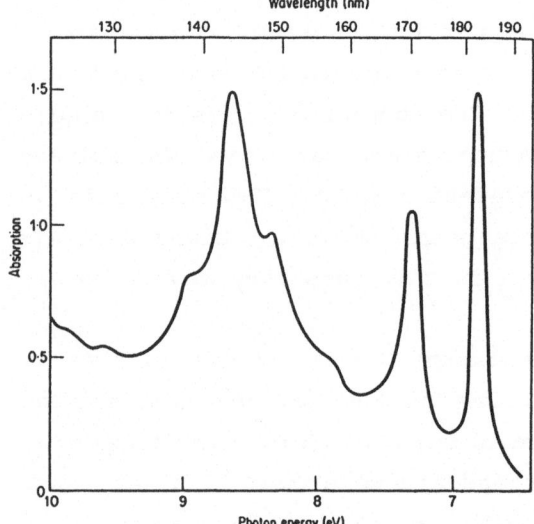

Fig.2.11. Absorption spectrum at 77 K of KBr near the absorption edge. The peaks around 6.8 and 7.4 eV are known to be exciton lines [2.93,97]

Generally, exciton features are not observable in materials of high dielectric constant and low effective mass, since their binding energy, $E_b \ll kT$, and their energy levels, are masked by the conduction band. Only at extremely low temperatures will their presence be noticeable. Furthermore, excitons are absent in materials containing a large density of electrons, such as metals and heavily doped semiconductors, since the free carriers screen out the field coupling the electron-hole pair.

At energies well beyond the band edge and the discrete excitation features, the absorption falls off rapidly to some value where generally the material absorption is at its lowest for the whole electromagnetic spectrum. This usually occurs for these materials when the wavelength is increased out into the near infrared. Throughout and beyond this point well into the microwave region, intraband electronic transitions become appreciable. In semiconductors as well as metals, where transitions can be confined to within the conduction band, or within the valence band, this phenomenon is known as free carrier absorption. It depends, amongst other things, upon the density of free carriers, their mobility, and the number of available states within the band the carriers occupy. In metals, the free carrier density is very high and in the infrared free-carrier absorption can dominate all other absorption features.

It has already been discussed in Sect.2.3.1 that the free electrons and holes can be successfully described as particles of a gas. An extension of these ideas leads to the renowned Drude model of free carrier absorption. By solving the classical equation of motion of a free carrier, with the sole quantum mechanical modification that the effective mass m^* rather than m_e is used, we may derive

$$2nk = \frac{\sigma_0}{(1 + \omega^2 t^2)\,\omega\epsilon_0} \tag{2.9}$$

$$n^2 - k^2 = \frac{\epsilon_L}{\epsilon_0} - \frac{\sigma_0 \omega t}{(1 + \omega^2 t^2)\omega\epsilon_0} \tag{2.10}$$

where $\sigma = Ne^2 t/m^*$, t is a relaxation time associated with randomization of the carrier velocity arising from collisions, N is the carrier density, and ϵ_L is the dielectric constant of the lattice. For specific conductivity approximate solutions to the above equations can be found.

For materials with low conductivity, such as semiconductors and insulators, we can approximate $n^2 - k^2 = \epsilon_1/\epsilon_0 = \epsilon_L/\epsilon_0$, in

43

other words, $\epsilon_1 \simeq \epsilon_L$, and the contribution of the electrons to the dielectric constant is negligible, and independent of frequency. Also for low conductivity, it is clear that $n^2 - k^2 > 2nk$, i.e. $n > k$, and therefore $n \simeq \epsilon_1/\epsilon_0$. Using these assumptions a value for the absorption coefficient due to free-carrier absorption can be derived as

$$\alpha = \frac{\sigma_0}{1 + \omega^2 t^2} \sqrt{\frac{\mu_0}{\epsilon_L}} \qquad (2.11)$$

where μ_0 is the permeability of free space. At low frequencies $1 + \omega^2 t^2 \to 1$ and the absorption is essentially frequency independent, but at high frequencies when $1 + \omega^2 t^2 \to \omega^2 t^2$, then it is proportional to the square of the wavelength. For many solids, $t \simeq 10^{-13}$ s and therefore t is close to unity at frequencies around 10^{12} Hz, which corresponds to a wavelength of 300 μm. At still higher frequencies, absorption by free carriers can no longer be considered as a simple energy transition by a carrier. The electron, for example, must also change its wave vector in making the transition and this must involve a change in crystal momentum e.g. by the lattice, or an impurity since at these energies the photon by itself cannot take up such large differences in momentum. It is known that at these near infrared wavelengths, the overall absorption can have a $\lambda^{1.5}$ or $\lambda^{2.5}$ dependence due to acoustic mode and optical mode scattering of the carriers, and also $\lambda^{3-3.5}$ dependencies arising from impurity scattering.

When the material conductivity is high, and $2nk \gg n^2 - k^2$, then for low frequencies ($\omega t \ll 1$), $n^2 \simeq k^2 \simeq \sigma_0/2\omega\epsilon_0$ and α is given by

$$\alpha = (2\sigma_0\omega\mu_0) \qquad (2.12)$$

The material exhibiting this characteristic under these conditions may be termed metallic. Inspection of (2.10) reveals that even in metals, when the frequency is sufficiently high, the free carrier contribution to the dielectric constant must

44

become small, and therefore (2.11) will be applicable. This occurs in the visible or UV for metals as discussed earlier in Sect.2.3.1.

Within the large background influence of free carrier absorption, there are often many strong features associated with the excitation of various discrete vibrational modes of the lattice (phonons). The absorption coefficients associated with this characteristic can be as high as 10^5 cm^{-1} in ionic solids, or more typically up to around 100 cm^{-1} for homopolar crystals such as silicon and germanium. This type of photon-phonon interaction occurs typically in the 0.1-0.01 eV region of the spectrum, as we have already seen for the case of optically induced vibrational excitation in isolated and absorbed gas phase species. Since the phonons have appreciable lifetimes and their wavefunctions are not localized, their energies and wave vectors are well defined as the usual energy and momentum conservation rules apply, although as we have previously discussed for free carrier absorption the wave vectors associated with photons can usually be neglected.

In ionic, or partially ionic materials such as GaAs, the motion of the ions gives rise to electric dipole coupling. The strong single photon absorption and subsequent interaction with a single optical phonon of wave vector approximately equal to zero occurs at the restrahl frequency. For pure covalent semiconductors, absorption arising from lattice vibration still occurs, though less strongly, and extends over a considerable spectral range. Because there is no electric dipole coupling in these cases, two phonon and higher order processes take place. Figure 2.12, for example, shows the absorption spectrum of single crystal silicon from 7-25 μm. Here the individual peaks have been assigned to various summations of at least two phonon energies, some of which involve previously identified modes arising from impurity effects, which will be discussed later. Momentum as well as energy can now be conserved during the interaction without introducing the need for individually small

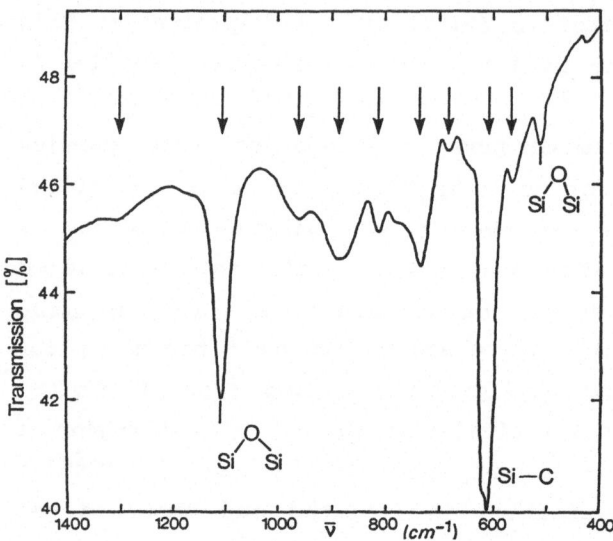

<u>Fig.2.12.</u> Infrared transmission spectrum of a silicon wafer as used in the microelectronics industry, indicating the various lattice and impurity absorption bands [2.94]

values of momentum for each phonon. In general the interaction can be written

$$E = h\nu = \hbar\omega_1(q_1) + \hbar\omega_2(-q_1) \quad \text{or} \tag{2.13}$$

$$E = h\nu = \hbar\omega_1(q_1) - \hbar\omega_2(-q_1) \tag{2.14}$$

In the latter case the phonon that is absorbed (minus sign) must be of significantly less energy than that created. Returning to Fig.2.12, with only one exception, all the lattice absorption bands of Si are less than $\simeq 1\text{cm}^{-1}$ in strength, and are found at 1302, 1107, 964, 890, 815, 740, 685 and 588 cm^{-1}; the more intense two-phon+on lattice vibration at 607 cm^{-1} exhibits an absorption coefficient of 10.3 cm^{-1}. Also present are absorption features due to lattice impurities unavoidably incorporated during the growth of the crystal. The periodicity of the lattice is broken by the presence of impurities, vacancies, defects etc., and this will, in fact, give rise to modified and even newly created bands. The presence of interstitial oxygen atoms in Si is characterized by the appearance of bands at 515 cm^{-1} and more obviously at 1106 cm^{-1} (spectral half-width 38 cm^{-1}), known to be due to vibrational modes of a nonlinear molecular unit of Si-O-Si resembling a water molecule. In addition, the absorption band at 602 cm^{-1}, which overlaps the 607 cm^{-1} feature mentioned earlier, is known to result from the presence of substitutional carbon atoms.

2.3.3 Nonlinear Optical Absorption

The optical absorption phenomena discussed up till now have for the most part been linearly controlled, i.e. the amount of absorption is proportional to the number of absorbing centres within the path of the radiation and the scattering cross-section. However, this linear absorption regime, characterized by the well-known Beer's law

$$I/I_0 = e^{-\alpha d} \tag{2.15}$$

can be disturbed in several ways. For example, if a new absorbing state is created once the absorption process has occurred, then the absorption can exhibit a nonlinear behaviour. This will happen when free carriers are generated across the bandgap of a semiconductor or insulator, which can subsequently contribute to enhanced absorption.

Multiphoton absorption processes can also occur, particularly in the presence of an extremely powerful laser beam, and of course the appropriate excitation in the material. In particular, the simultaneous absorption of two photons is a common event in semiconductors. For excitation of valence electrons to the conduction band via two-photon absorption, energy conservation requires

$$E_g \leq h\nu_1 + h\nu_2 . \tag{2.16}$$

In the more common cases where $\nu_1 = \nu_2$, the probability of the occurrence of two-photon absorption (TPA), being proportional to the probability of the availability of both photons, is proportional to the square of the intensity of the laser beam.

Free carrier absorption (FCA) is similar in form to TPA, except that the time delay between the two participating photons can be as long as the lifetime of the photogenerated free carrier in the excited state, which may be as long as 100's of nanoseconds. Hence, for pulses shorter than the carrier lifetime, while TPA is always intensity dependent, FCA depends only upon the number of free carriers created, and is approximately fluence dependent.

2.3.4 Plasma Formation

The extremely high intensity and ultrashort duration laser pulses that are readily available from many laser systems, have been known over the years to induce the formation of dense plasma above the surface of many materials. Indeed plasma formation by laser-induced dielectric breakdown at a gas-solid interface has attracted considerable attention over the years [2.12,13]. This process of laser induced optical breakdown involves the radiative heating of free electrons within the solid by the intense laser pulse. These initial electrons are thought to be present because of either thermionic emission, or some tunnelling process [2.14]. The electrons couple very strongly to the incident radiation, and by absorbing the energy are strongly accelerated, and collide with ions and various gas-phase atoms (inverse bremsstrahlung). In this way more atoms are ionized and they in turn provide more electrons which further increase the rate of ionization. This rapid formation of an electron gas by impact (sometimes known as cascade ionization) results in the remaining energy in the laser beam being absorbed above the surface of the material, thereby leaving the sample relatively cool, and shielded. Therefore, in an oxygen ambient, heterogeneous chemistry can be initiated by the laser beam. This technique has been used to induce the growth of thin oxide layers on the surface of niobium films [2.15], as will be discussed later in Chap.6, and has been designated laser pulsed plasma chemistry (LPPC).

2.4 Interactions with Surfaces and Adsorbates

This section is distinct from the text on laser interaction with solids because of the special properties of surfaces that make them difficult to compare with bulk materials. Additionally, it is most often the surface that serves as the rendezvous for the gas phase species, and the site of many of the subsequent chemical reactions. Often it is the special properties of the surface that actually induce or enhance a particular mechanism. In fact there are many ways in which the gas–solid interface can affect a reaction that, for example, another solid-solid interface, or a bulk material cannot. In this section the mechanisms behind optically induced desorption and adsorption, and the various modes of optical excitation of the adsorbate–surface complex will be discussed.

2.4.1 Adsorbates

When an atom or molecule interacts with a clean surface, it inevitably leads to a perturbation of the electronic structure of the surface region, which most likely already possessed a significantly different structure to the bulk material because of the truncation of the atomic lattice. If this adsorbate remains relatively inert such that the attractive atomic interaction is very weak, then the species is physisorbed. On the other hand, if the foreign material is reactive and a new chemical bond is formed, then the species is said to be chemisorbed. While both types of adsorption induce a change in the electrostatic potential of the surface, only the chemisorption influences the valence levels of the substrate. In fact a completely new set of modified electronic energy levels now results locally within the surface and the adsorbed species complex. Although adsorption can occur under a wide variety of conditions, the process can be significantly influenced by the presence of laser radiation.

Since chemisorption is highly specific in character, depending on the individual properties of the gas adsorbate and surface adsorbent, it is usually expected that only single monolayers (and no more) can be formed in this way. However, since physisorption involves van der Waal-type forces, although these are significantly weaker, they can extend from one monolayer to another. It can be anticipated, therefore, that several monolayers can under particular circumstances be physisorbed onto a surface.

Once adsorption has occurred it presents an additional set of possibilities for laser assisted surface interactions. Photochemical coupling of the radiation could now occur where before it was not possible because of the lack of appropriate absorption sites. Consequently, a new avenue is now available for deposition, etching, and film growth.

The interaction of laser radiation with even the lightly contaminated surface, however, can also give rise to the complimentary process of desorption. Here, the laser radiation directly, or indirectly, encourages bond-breaking between adsorbate and surface, and consequently the impurity leaves the surface.

We have already seen for many molecules, that consecutive excitations up the vibrational ladder cannot be practically achieved using lasers because of the usual high degree of anharmonicity (Sect.2.2.1). For example, to reach the n'th level of vibrational excitation, n different photons of specific frequency would be required. On the other hand, when these molecules are adsorbed on to a surface, the number of available excitational modes is increased significantly. For example, in isolation a diatomic molecule has only one single mode of vibration. Upon adsorption, it can have up to six modes of vibration, including hindered translational and rotational motions. As with polyatomic molecules described earlier, this increase in the density of states brings the level of the quasi-continuum lower down into the Morse potential and therefore makes it more easily accessible to smaller numbers of multiple photon absorption.

The vibrational activation of adsorbate–surface complexes is damped rapidly by the surface itself. Nevertheless, it should be possible to excite other intramolecular vibrational modes during this relaxation, such that, before the energy is transferred to bulk phonons and manifested as heat, indirect excitation of a weaker bonding unit can occur. Clearly the various time scales involved must be carefully summed up and compared with competing processes, and consequently it is often quite difficult to identify the more complex polyatomic intermolecular energy transfer pathways. Figure 2.13 gives a good picture of an intramolecular energy transfer mechanism by which excitation of a mode not even directly connected with the adsorbate vibrational modes results in the eventual excitation of the modes after some period of time [2.16]. This can often lead to desorption of a species from the surface, as will be discussed in the following subsection.

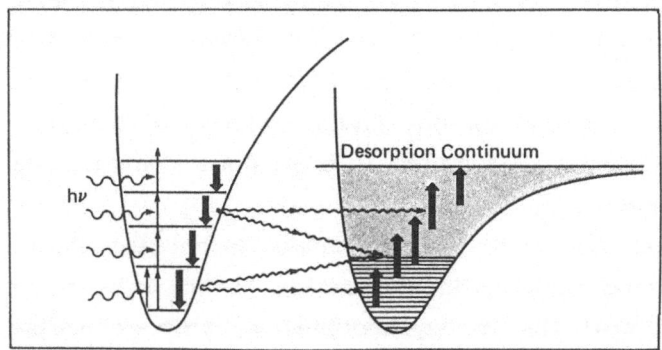

Fig.2.13. Intramolecular energy transfer from the excited high frequency photoactive mode to the low frequency adsorbate surface stretching mode may cause desorption of the adsorbate species [2.16]

2.4.2 Desorption

As with most laser-assisted processes, desorption can be governed by the changes in the thermal properties of the material resulting from laser heating, or it can be initiated by any one of several photoinduced absorption mechanisms. In laser-assisted thermal desorption, the foreign species is released from the surface when energy is transferred by the energetic phonons from the hot substrate into the surface–adsorbate bond. The molecular desorption occurs with a characteristic translational temperature of the desorbed species. In this case the laser radiation is absorbed by the substrate and the modified electronic structure does not play a significant role in the process. Laser–induced evaporation will be discussed later in the section.

Photodesorption can occur in several ways. As with gas phase species, the radiation can be coupled into either the vibrational modes of the system or to electronic transitions. Again, as expected, when the energy is resonantly absorbed into the internal vibrations of the adsorbate–surface system, the energies involved are associated with the infrared region of the spectrum. It is thought that the laser–assisted desorption of CH_3F from NaCl, recently investigated using a CO_2 laser, is a multiphoton excitation process [2.17]. Similarly, pyridine molecules adsorbed on crystalline KCl [2.18] or on silver on SiO_2 [2.19] can be desorbed by CO_2 laser pulses, although in this case it is suspected that there is also some contribution from thermal processes.

In the ultraviolet region of the spectrum, desorption can be initiated by promoting electronic excitation of the system. It had been suggested that the energy absorbed by this mechanism may not remain localized for sufficiently long times within the excited adsorbate–substrate complex. Consequently, thermal redistribution takes place, and the dominating desorption process will be mainly thermal [2.20]. This phenomenon has attracted a considerable amount of interest in recent years [2.20-22] and in the UV region desorption of both neutral and ionic species has been observed [2.23,24]. Basically, the transitional excitation to a repulsive neutral or ionic state occurs, via $M + A + h\nu \rightarrow (M + A)^*$ which in the case of neutral desorption becomes $M^* + A$, and in the case of ionic desorption

52

becomes $M^- + A^+$, where M is the (metal) surface and A is the adsorbate. If A is a molecule rather than an atom, similar mechanisms can produce photofragmentation. Multiphoton electronic excitation has also been recently identified as another mechanism by which both ionic and neutral species can be desorbed [2.22]. The study of CO on Ru, for example, was found to give rise to CO^+, O^+, and CO, and a detailed analysis of the thresholds of the various species was reasonably interpreted in support of this mechanism.

The dynamics by which species are photodesorbed physically from surfaces is not yet clearly understood, and require more fundamental investigation. SCHAFER and HESS [2.25] recently reported time-of-flight (TOF) diagnostics of CO_2 laser-induced desorption of, e.g. CH_3F, CH_3COOH, and CCl_4 on various substrates such as SiO_2, and conclude that the observed wavelength-dependent photodesorption may be adequately described by resonant heating via selective excitation of the internal modes of the condensed molecules. Figure 2.14 shows the tem-

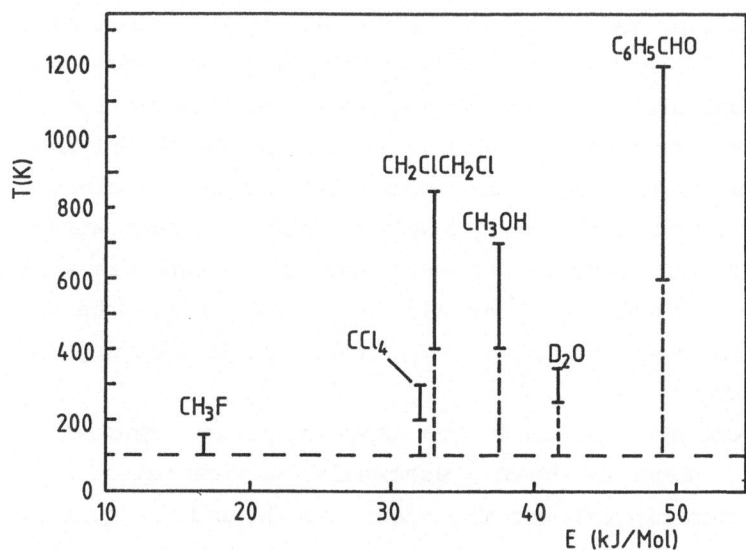

<u>Fig.2.14.</u> The range of temperatures determined by time-of-flight measurements plotted against the heat of evaporation for various compounds [2.25]

peratures determined experimentally for different molecules versus the heat of vaporization. The good correlation indicates that the heat of vaporization can be taken as crude measure of the strength of the intermolecular interaction in the van der Waals films. At present, for metal surfaces several possible mechanisms exist that are quite feasible though not yet associated with any particular observations. For example, one such model [2.26,27] involves an optically induced excitation causing a Franck-Condon transition of the adsorbate-substrate complex from the ground state to a neutral or ionic antibonding state, whereupon the particle would begin to move away from the surface. Clearly the physical departure of the species from the neighbourhood of the surface will take a finite amount of time. During this "escape-time", if de-excitation occurs, the particle would once more become adsorbed on the surface. Such a process is always possible, especially if an electron could tunnel across to fill the orbital emptied by the initial excitation.

Since the migration of reactants adsorbed on the sample surface can be an important step in the overall rate process of a surface chemical reaction, SLUTSKY and GEORGE [2.28] have considered the possibility of laser stimulated surface motion of adsorbed atoms. Their theoretical approach has predicted that laser radiation could be used to enhance diffusional-type behaviour under conditions of steady state excitation. Laser stimulation of electron migration between surface groups has also been suggested by DZHIDZHOEV et al. [2.29] to explain the increased rate of decomposition of NH_2 groups on SiO_2 during CO_2 laser irradiation.

Another proposed mechanism for desorption is depicted in Fig.2.15 [2.30]. Here is shown a schematic representation of a probable relationship between the adsorbate (A) and the surface (usually represented by M, strictly for metal surfaces, but here we generalize the M to mean any surface), in terms of potential energy and physical separation distance. The potential

54

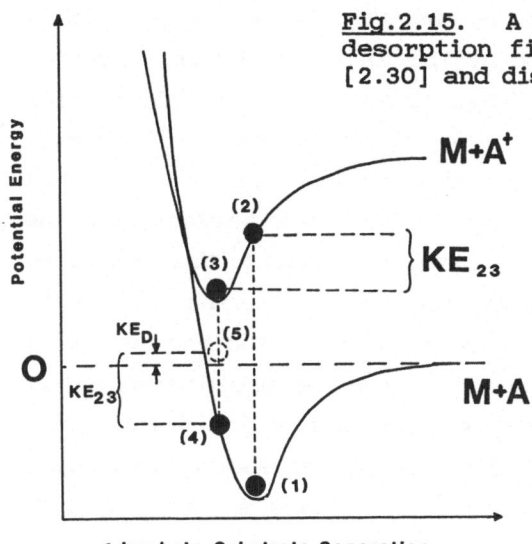

Fig.2.15. A possible mechanism for desorption first proposed by ANTONIEWICZ [2.30] and discussed by CHUANG [2.16]

energy curves indicate the total energy of the system when the adsorbate is in the neutral state (A) or in the ionic state (A^+). In this model it is important to note that the species is substantially smaller in its ionic state than in its neutral state, such that the potential of the excited system had its minimum at smaller atomic separation than did the ground state. Upon excitation from $M + A$ to $M + A^+$, the adsorbate is accelerated towards the solid surface gaining kinetic energy. Nearer the substrate the probability of electron tunnelling was increased and inevitably the ionic species was neutralized. Following de-excitation to the lower curve, the system entered into an immediate state of repulsion. If the sum of the potential energy of repulsion and the newly acquired kinetic energy is greater than the binding energy of the M+A system then the species is desorbed from the surface. Similar arguments for desorption of ionic species involving three potential energy curves have also been proposed.

Desorption pathways involving chemical reactions as essential components of the multistep process, when the systems were exposed to ultraviolet-visible radiation, have been proposed for desorption from oxide and sulphide surfaces [2.31]. Neutral CO_2 has been found to be the major or the only species desorbed from ZnO, TiO_2, and V_2O_5, with quantum photodesorption yields as high as 10^{-3} being sometimes achievable [2.32-34]. The essential requirements for this to occur were the presence of carbon impurities in the region of the oxide surface, and the presence of oxygen in the immediate environment. This resulted in the creation of a CO_2 species, which included the capture of conduction band electrons from the substrate, and the neutralization to CO_2 by recombination with the holes photogenerated at the surface. The threshold for this process, which required the presence of optically generated carriers, was shown to be equal to the bandgap energy for the various materials involved (2.7, 3.1 and 2.35 eV, respectively).

In more detail, VAN HIEU and LICHTMAN [2.33] found that there was no desorption of CO_2 when radiation whose frequency was more than 400 nm was used. Further into the UV, the CO_2 evolution from the surface exhibited a linear relationship with the intensity of the radiation incident on the substrate, and it was also found that the desorption rate was inversely proportional to $t^{1/2}$ (where t is the irradiation time). Whereas this group used a mercury lamp as the source of the visible-UV light, KAWAI and SAKATA [2.35] employed a pulsed nitrogen laser to irradiate both ZnO crystals, and TiO_2 powders suspended in water and coated onto TiO_2 crystals, after exposure to air. With fluences less than 3 mJ/cm^2, both CO_2 and H_2O were observed. The roles of the hydroxyl groups or indeed the carbonates on the surface are not fully understood at present. With increased fluences, both ionic species and molecules were desorbed.

Surface roughness can of course give rise to electromagnetic fields many orders of magnitude greater than the incident field, as well as periodic variations in the optical and electronic properties of the irradiated material. Although not yet identified in any desorption system, such effects could conceivably become important under the appropriate conditions.

2.4.3 Adsorption

As outlined above, various species can be adsorbed on to a surface, either in a weak physical sense involving van der Waals-type bonding, or in a true chemical sense involving a sharing of electrons. The presence of optical radiation can strongly influence the behaviour of the gas–surface system in a number of ways. Indeed, these optically induced interactions contribute significantly to many of the film-forming and microstructural characteristics described in some of the following sections of this book. The reader is also referred to recent reviews by CHUANG [2.36,37]. Nevertheless, although the mechanisms involved are amongst the more important aspects of these laser-based techniques, they are, unlike the much more widely studied homogeneous chemical reactions, not very well understood.

It is most likely that adsorption occurs at all surfaces, although historically it has been identified in porous substances that exhibit large surface areas, the best known of which are charcoal, silica gel, aluminium oxide, chromium sesquioxide (Cr_2O_3) and ZnO. Although a large surface area is quite clearly important for adsorption, there are many other very important mechanisms related to particular properties existing between the adsorbate and the surface that determine the extent of adsorption. For example, it is well known that in general an increase in pressure and a decrease of temperature increases the extent of adsorption. This relationship is usually described by adsorption isotherms which show the amount of adsorption over a range of pressures at a specific temperature. Over a limited range of pressure, at constant temperature, the actual variation of adsorption with pressure P can usually be represented by the empirical relationship

$$a = mP^n \tag{2.17}$$

where m and n are constants for a given surface and adsorbate combination at a particular temperature, and a is strictly the amount of gas adsorbed per unit mass of adsorbent. A more accurate relationship, known as the Langmuir adsorption isotherm, relating a to P for a given temperature is given by

$$a = k_1 P/(k_2 P + 1) \qquad (2.18)$$

where k_1 and k_2 are constants. This formula is strictly only applicable to monolayer coverage and numerous instances of this relationship have been reported over the years, understandably, prime examples of chemisorption.

At very low temperatures, the value of $k_2 P$ becomes negligible compared to unity, and there exists a special case of the adsorption being proportional to the pressure. For high pressures, unity becomes insignificant compared with $k_2 P$, and the adsorption is determined by the ratio of k_1/k_2. This limiting adsorption with increasing pressure would be applicable at any temperature, and is a consequence of one monolayer coverage on the surface. At intermediate pressures, expressions of the form of (2.17) become applicable with quite reasonable accuracy.

Although physical adsorption can also result in only one layer coverage of the surface (Type I), it is usual to find a multilayered structure built up on the adsorbent. In fact, five types of physisorption isotherms have so far been identified. By applying the Langmuir approach to such multilayered physisorption, BRUNAUER, EMMETT, and TELLER (BET) derived the equation [2.38]

$$\frac{v}{v_m} = \frac{PP_0}{P_0 - P} \cdot \frac{c}{P_0 + P(c-1)} \qquad (2.19)$$

where P_0 is the pressure required to condense the gas at the operating temperature (i.e., the vapour pressure), v is the volume of gas adsorbed at pressure P, v_m is the volume adsorbed when the adsorbent is covered by a single layer of molecules, and c is a constant approximately equal to $\exp(\delta E/RT)$, where δE

58

is the energy associated with the adsorption of the first mono-layer. When $c > 1$ and $c < 1$, adsorption characterized as type II, and type III, respectively, is achieved. Types IV and V occur as modifications to types II and III if gas condenses inside some of the small pores and capillaries of the surface struc-ture at pressures below P_0. From a plot of $P/v(P_0-P)$ against P/P_0 the slope of the straight line isotherms should give according to the BET equation above (2.19), the ratio of $(c-1)/v_m c$, and the intercept $1/v_m c$. From these ratios, the value of v_m can be calculated, from which, knowing the molecular diameters and assuming close-packing, the area of the adsorbent surface can be estimated. It is important to note that the ratio v/v_m is sometimes defined as θ, the fractional surface coverage. Recent papers by CHEN and OSGOOD [2.39], and EHRLICH and OSGOOD [2.40] describe several aspects of physisorbed molecules, particularly metalalkyls, in relation to photochem-ical deposition of metallic layers.

As an example of oxygen adsorption, Fig.2.16 shows chemi-sorbed oxygen on Ni for increasing exposure at two different temperatures [2.41]. It appears that the 77 K behaviour is indi-cative of type I described above, while the 300 K characteristic is more representative of type IV. Figure 2.16 also shows the amount of oxygen chemisorbed on cleaved GaAs surfaces held at 120 and 300 K [2.42]. Interestingly, the insert in Fig.2.16(b) indicates that if the system is evacuated after initial oxygen uptake at 120 K, then some previously adsorbed oxygen will desorb with time, eventually resulting in a new equilibrium uptake. Although the initial sticking coefficient is about 40 times greater at 120 than at 300 K for small exposures, oxygen coverage at large exposures shows a stronger tendency to satu-rate around half a monolayer at 120 K.

Simultaneous exposure of the GaAs to xenon arc lamp radia-tion and $4.5 \cdot 10^6$ L of oxygen is found to more than double the coverage that is obtained in the dark under these conditions. Although the temperature rise is not reported in their work

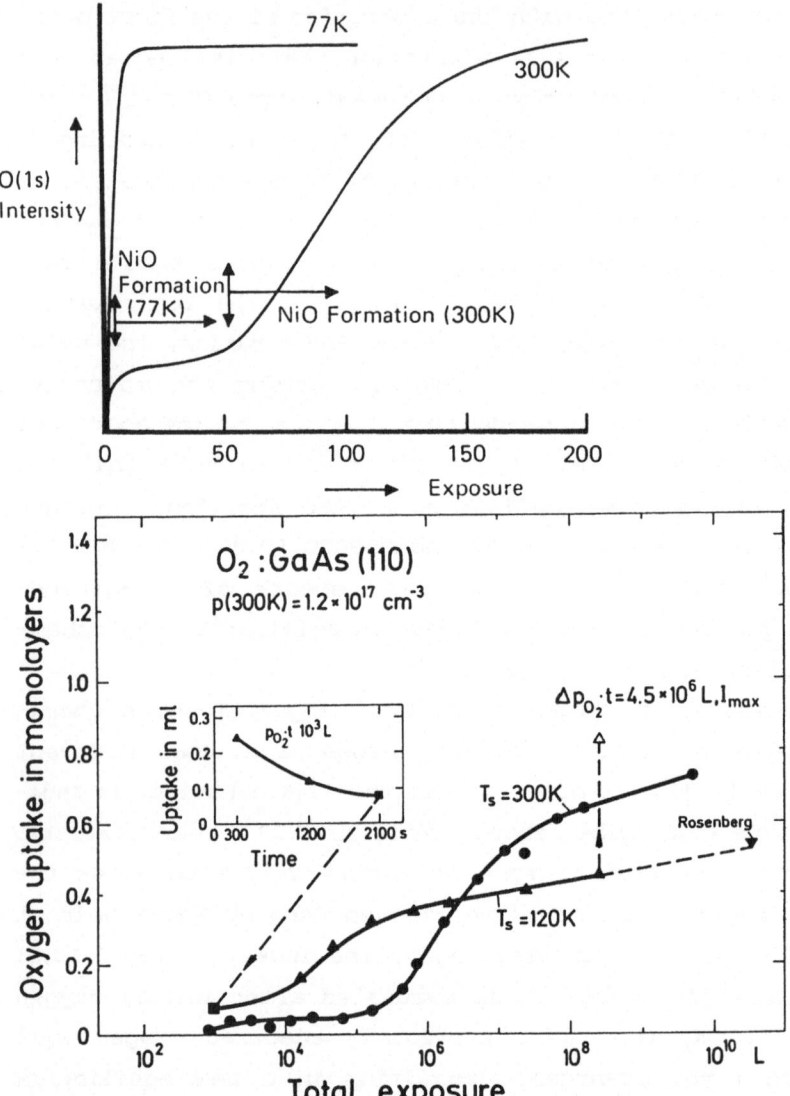

Fig.2.16. (a) Chemisorbed oxygen uptake as determined from the O(1s) intensity of XPS data, versus exposure at 77 K and at 300 K, for Ni(100), by BRUNDLE and HOPSTER [2.41]. (b) Uptake of oxygen on cleaved GaAs surface at 120 and 300 K (see text for detail) [2.42]

[2.42], BARTELS and MÖNCH have proposed that the photogener-ated electron-hole pairs are instrumental in stimulating appre-ciably more oxygen uptake than is otherwise possible in the dark at much higher exposures. The influence of optical radiation on adsorption as well as the subsequent oxidation chemistry on GaAs is briefly reviewed later in Sect.4.5. The coverage of a surface with an adlayer can be achieved using a variety of optical techniques. These include vibrational or electronic excitation of the gas phase species or any adsorbate–substrate complex already formed, excitation of carriers across the bandgap of the material itself, dissociation of the gas phase molecules, and any thermal processes that occur as a result of optical absorption within the substrate. The substrate–related mechanisms will be discussed in Sect.2.6.

Interactions of laser radiation with the gas phase have already been addressed in Sect.2.2, where the concept of vibrational and electronic excitation was introduced. Although the importance of such photo–excited species in many reactions is unquestionable, little is known either experimentally or theoretically, about the precise reaction pathways involved in many such cases, particularly where surfaces play an influential role, and consequently much remains to be investigated in this regime.

If bond dissociation occurs, the photocreated species may then interact with a surface that was previously inert to the undisturbed parent molecule. Small amounts of energy may be released upon adsorption of these species, while any ensuing exothermic chemical reactions will contribute even greater thermal energy to the system, introducing further complications to identifying the dominant reaction mechanisms. Multi-component reactions influenced by surface adsorption present greater analytical difficulties. When metal oxides, for example, are photodeposited from alkyl or carbonyl derivatives and oxygen–bearing donor molecules, it has been established that the reaction depends quite strongly on the irradiation geometry and laser wavelength, as well as the molecular species and the usual ambient considerations. This will be discussed later in Sects. 6.1,2 which are concerned with the photodeposition of oxide layers.

61

The influence of photogenerated electron-hole pairs on various chemical reactions is often examined in optically initiated chemical reactions on surfaces. When the electron-hole pairs are created by photonic absorption they can recombine within their natural lifetime and produce heat, or they may diffuse to the sample surface where they may either produce heat, or contribute to a chemical reaction, by either the electrons or the holes being captured by the reaction elements. For example, an oxygen molecule can be photoadsorbed by capturing an electron or photodesorbed by capturing a hole such that

$$e^- + O_2(g) \rightarrow O_2^- \text{ (ads)} \tag{2.20a}$$
$$h^+ + O_2^-(\text{ads}) \rightarrow O_2(g) \ . \tag{2.20b}$$

The likelihood of either reaction is strongly dependent upon the detailed electronic structure of the irradiated system.

We now know that isolated molecules dramatically increase the number of vibrational modes by being adsorbed on to a surface, and this broadening and the spectral shifts together with the high molecular densities possible with adlayers enable subsequent film nucleation to be precisely controlled by direct photolysis of the surface adsorbed species. In this way ultrahigh resolution is possible, because of the lack of importance of particle diffusion in the gas phase, or thermal diffusion in the solid phase. Once a few nucleation sites have been established, deposition can be continued via normal gas-phase pathways. Alternatively, under the most appropriate conditions, deposition can continue in this adsorbed-layer mode with only a negligible contribution from the gas phase. For example, EHRLICH and TSAO [1.43] have reported that successively thicker layers of $Al_2(CH_3)_6$, in the 1-10 monolayer regime, have been condensed on to SiO_2 by slight cooling of the substrate to below room temperature.

A final example is one of a surface catalysed process in which adsorbed layers of the organic molecule methyl methacrylate (MMA) are polymerized into films of poly-MMA, or PMMA, by

62

absorption of UV laser radiation [2.44]. Molecules of MMA from the ambient form on the surfaces exposed to the vapour and the free radical catalysed polymerization is initiated while rapid collisions of further vapour-phase species continually replenish the adlayer. It was found that the polymer growth, initiated by a weak photodissociative absorption at 257.2 nm, could be increased substantially using a compound adsorbed layer structure. This sensitizing layer was formed by exposing the original SiO_2 substrates to $Cd(CH_3)_2$ vapour, then evacuating the cell, and introducing the MMA. Although the technique leaves a single chemisorbed monolayer under the PMMA, an enhanced preparation rate of nearly two orders of magnitude (up to 1 $\mu m/s$) can now be achieved. More recent work by the same group shows that UV laser photochemical modification of adsorbed $TiCl_4$ and $Al(CH_3)_3$ on quartz substrates produces highly optimized catalysts for in situ patterned polymerization of ethylene and acetylene [2.44].

As the general level of materials technology increases, and surface structures are continually required to be more complex and smaller in size, yet more controllable both in manufacture and application, the importance of surface structure and surface quality will become more evident. Surface preparation for VLSI devices is already an important and well established area where the use of lasers to promote desirable conditions is being seriously investigated. Just as the adsorption processes are important for the deposition of thin films and microstructures, so too can be the desorption mechanism. We have seen that sometimes adsorption of particular species is a prerequisite before desorption can occur. Conversely, it is not difficult to imagine a situation where desorption must take place before adsorption or some other process can be initiated. Although most of the work in this area of laser processing is very much application-driven, by both demand and funding, it is of considerable and continuing benefit that there remains a healthy degree of fundamental interest in the mechanisms involved. As we have seen in this subsection, the interaction of gaseous species and surfaces is an extremely complex situation, and the presence of energetic photons, although sometimes providing an important clue towards the understanding of some process, can sometimes make it even more difficult to interpret the basic mechanisms involved. However, if the full potential of laser assisted processing of thin layers is to be realised, then a basic understanding of all of these surface-related phenomena is necessary.

2.5 Laser Induced Heating

2.5.1 Thermalization

With the exception of direct excitation of the lattice vibrations of the material, all the other absorption processes discussed above involve electronic transitions within the energy bands of the solid. In these cases none of the interactions therefore directly activates lattice vibrations, inducing the atoms into some form of motion thereby heating up the material. The absorbed energy does however in most cases increase the the density of free electrons and holes, and also provides them with a significant amount of kinetic and potential energy. Once energy has been coupled into the carrier system, it must eventually be redistributed amongst the other carriers through carrier-carrier interactions, between carriers and lattice through carrier-phonon scattering, and also within the lattice through phonon-phonon interactions. Ultimately, the material may heat up significantly in extremely short times.

Within the electronic system itself, these rapid exchanges of energy occur by several mechanisms, whose time scales are plotted in Fig.2.17 [2.45]. The scattering processes involve carrier-carrier and carrier-plasmon (an oscillation of the carrier plasma) collisions, and these depend upon the actual density of the carriers. As shown in the figure, these are extremely rapid, occurring in times of the order of 10 fs for

64

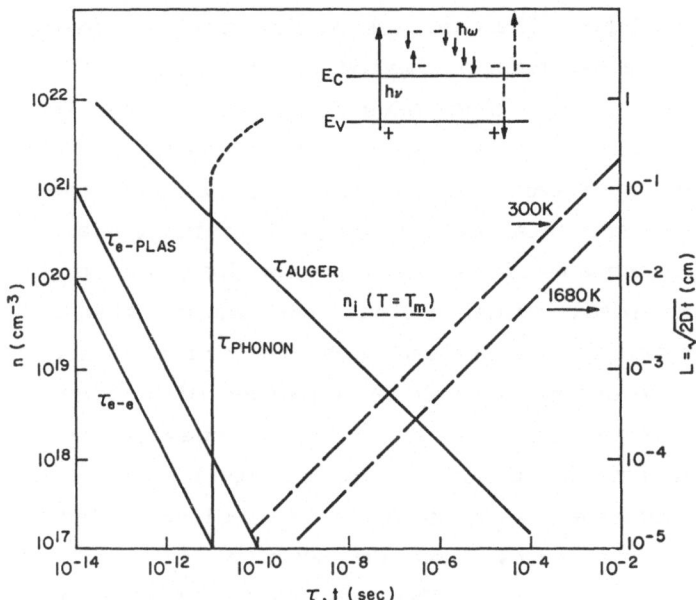

Fig.2.17. A summary of the various electron scattering times in silicon, which also applies in general to most other semiconductors, and in some instances to metals and insulators. Also included are the characteristic heat diffusion lengths for silicon at two temperatures in the solid phase. The insert is a schematic of some of the important scattering processes. The vertical axes show the carrier density on the right, and the thermal diffusion length on the left [2.45]

densities around 10^{20} carriers per cm^3. Another process which assists in the redistribution of energy within the carrier system is the Auger process. Here an electron and hole recombine and transfer their recombination energy to a third carrier. In this way the density of the plasma can be significantly reduced, although the total energy will remain approximately constant within the carrier system. (The opposite mechanism of impact ionization can also occur by the creation of an electron hole pair by a sufficiently energetic third carrier).

Energy can escape from this quasi-thermalized system via carrier-phonon scattering, and in this way the lattice will heat up according to the number of lattice phonons created. Typically, about 20 phonons must be generated in order that 1 eV can be transferred from the carriers to the lattice, and the time required for this has been estimated to be less than 10 ps [2.46] for $n \simeq 10^{21}$ cm^{-3}. There have been predictions that screening effects will block the transfer of energy in this way when larger densities of carriers are created, although there are no conclusive experimental data in support of these claims. Indeed there is a consensus of opinion from the large amount of accumulated experimental evidence that the energy transfer time from the carrier to the lattice is of the order of a few picoseconds. Since we are not presently concerned with times of the order of picoseconds for these laser induced chemical reactions, it is appropriate to consider that the lattice is heated essentially instantaneously as a consequence of direct irradiation by intense laser beams from cw down to time scales approaching the picosecond range. The consequence of such rapid heating, and subsequent cooling phenomena is that a completely novel and unique technique of thin film preparation is now available, that can give rise to a wide range of new types of layers with several advantages inherent in the processing technique itself.

2.5.2 Heating Models

A great many models have been used to successfully explain and predict the effects produced by laser induced heating and melting of materials. Although often described as "simple" thermal models, these are often quite complicated, involving nonlinear optical and thermodynamical processes such as absorption, reflection, carrier and thermal diffusion, melting and evaporation, energy of reaction, and indeed the reaction mechanism itself. Not surprisingly, then, there does not exist a universal model describing laser induced heating, and in general, many groups have understandably only modelled the situation for a small range of materials whose optical and thermal properties are similarly described, in terms of wavelength and temperature dependencies. In many cases, however, several simplifying assumptions with little loss in accuracy, can be used in order to provide analytical rather than numerical solutions to each problem. Often, all constants are given an average value throughout the calculation if they do not vary by much during the heating cycle, with perhaps only the strongest nonlinearity being given some (usually empirical) effective nonlinear dependence on the calculation.

It is instructive to study a few illustrative cases which may operate under certain heating conditions. For example, in cases where the heating cycle is short, and beam dimensions are in general quite small, there is usually no significant contribution to heat loss from the substrate arising from blackbody re-emission of the absorbed energy. Similarly, there is rarely a loss of energy from convective currents in the surroundings. Therefore many heating models although there are notable exceptions, are based upon variations of the well-known equation of heat diffusion. Here the induced temperature distribution T, strictly T (x,y,z,t), arising from a laser beam propagating along the z axis and impinging upon the sample surface which is uniform within the x,y plane can be described by

$$\frac{\delta T}{\delta t} = \frac{Q}{\rho C_p} + \frac{\delta}{\delta z} \left(\frac{D \delta T}{\delta z} \right) \tag{2.21}$$

where Q, strictly Q (x,y,z,t), is the source term arising from the Gaussian laser beam having been absorbed within the material and D, the thermal diffusivity, is given by $D = K/C_p \rho$. In a general form, Q is written

$$Q = I(r,t) \cdot (1-R) \cdot F(z) \tag{2.22}$$

where $I(r,t)$ is the incident beam intensity, R is the sample reflectivity, and F(z) is a function that expresses the penetration depth of the laser light in the material, while K, C_p, and ρ represent the thermal conductivity, specific heat and density respectively.

67

Before considering more detailed solutions of (2.21), it is instructive to illustrate two simple and extreme cases of laser induced heating [2.47]. Firstly, and throughout the initial stages of this section, it is assumed that the workpiece is semi-infinite, and that all constants used are independent of temperature. If we assume that the laser beam is uniform in the plane of the solid, and that the material is a homogeneously absorbing medium, Q can be rewritten

$$Q = I_0(t) \cdot (1-R) \, \alpha \, e^{-\alpha z} \qquad (2.23)$$

where I_0 is the output power density from the laser, and α is the linear absorption coefficient of the solid. Using this value of heat source, (2.21) can, in fact, be solved analytically under the conditions already outlined. The reader is referred to CARSLAW and JAEGER [2.48] for further details on the calculation, from which the exact solution is obtained

$$T(z,t) = [(\frac{2I_0}{K})\sqrt{Dt} \; \text{ierfc}(\frac{z}{2\sqrt{Dt}}) - (\frac{I_0}{\alpha K}) \, e^{-\alpha z}$$

$$+ (\frac{I_0}{2\alpha K}) \, e^{\alpha^2 Dt - \alpha z} \, \text{erfc}(\alpha\sqrt{Dt} - z/2\sqrt{Dt})$$

$$+ (\frac{I_0}{2\alpha K}) \, e^{\alpha^2 Dt + \alpha z} \, \text{erfc}(\alpha\sqrt{Dt})](1-R) \; . \qquad (2.24)$$

At this point, the two extremes can be considered [2.47]. Firstly, it must be recognised that the heating process can in general be described in terms of two characteristic lengths, namely α^{-1}, the optical absorption depth, and L_d, the heat diffusion length, which is approximately equal to $(2Dt)^{1/2}$, i.e. $(2Kt/C_p)^{1/2}$. In cases where L_d is much greater than the optical absorption depth α^{-1}, as it is for most metals at most laser wavelengths, or for semiconductors and insulators at wavelengths above their bandgaps, the heat source term becomes a surface source of energy, since z becomes very small and α becomes large. As a consequence, (2.24) is reduced to

68

$$T(z,t) = \left[\left(\frac{2I_0}{K\sqrt{Dt}} \right) \, \text{ierfc} \left(\frac{z}{2\sqrt{Dt}} \right) \right] (1-R) \tag{2.25}$$

and if we concentrate on the temperature increase at the sample surface (z=0), then

$$T(0,t) = \left(\frac{2P(1-R)}{\pi K r_0{}^2} \right) \sqrt{\frac{Dt_p}{\pi}} \; . \tag{2.26}$$

Hence the incident power required to bring the surface to a particular temperature is proportional to the square root of the duration of the incident pulse since energy diffuses appreciably into the lattice as it is being deposited at the surface of the workpiece.

In the other extreme, when $L_d < \alpha^{-1}$, the temperature rise induced by short pulses is no longer controlled by the diffusion of heat since the energy is absorbed over greater depths than can be immediately influenced by thermal diffusion. In this case the temperature rise as a function of depth is given by

$$T(z) = P(1-R) \, \alpha \, e^{-\alpha z} \, \frac{t_p}{\pi \rho C_p r_0{}^2} \; . \tag{2.27}$$

Therefore the heating rate, given by $T(z)/t$ is independent of the duration of the laser pulse, and decreases exponentially with depth as expected. These equations are useful under the conditions of a homogeneous beam, and temperature independent optical and thermal properties. It can be seen that in both instances the increase in temperature in the sample is dependent upon the intensity of the incident laser beam, as would probably be expected.

In practical situations, however, the spatial profile of the laser beam is not homogeneous, but rather, is described more commonly by a Gaussian spatial profile. Under these conditions, the energy source term for the diffusion equation can be generally written

$$Q = \frac{P(1-R)}{r_x r_y} \ F(z) \ \exp\left[-(x^2/r_x + y^2/r_y)\right] \qquad (2.28)$$

where r_x and r_y define the major and minor axes of the elliptical beam. If the beam is scanned across the workpiece in say
the x direction, then the parameter x can be replaced by (x-vt)
where v is the scanning velocity. The use of (x-vt) in place of
x is strictly only valid in cases where cw scanned beams moving
across the sample surface lead to dwell-times (the actual time
for which one particular spot on the sample is exposed to the
moving beam, $t_{dwell} \simeq r/v$) that are much longer than the thermal time constant of the material ($t_{thermal} \simeq r^2/2D$), i.e.

$$v \ll 2D/r \ . \qquad (2.29)$$

Under these conditions the sample can be considered to have
had sufficient time to come into thermal equilibrium with the
laser beam before it moves on. For simplicity, however, only
stationary laser beams will be considered in this analysis.
Also, for these illustrative cases, only circularly symmetric
beams will be considered i.e. $r_x = r_y = r$. After integrations
over x, y, and z, the temperature increase can be written

$$T(0,0,0,t) = \frac{2P(1-R)}{K\pi^{3/2}r} \ \tan^{-1}\sqrt{\frac{4Dt}{r^2}} \ . \qquad (2.30)$$

Therefore the maximum temperature rise is given by

$$T(0,0,0,t) = T_{max} = \frac{P(1-R)}{2Kr\sqrt{\pi}} \ . \qquad (2.31)$$

In this case, therefore, with the assumption that R and K are
temperature independent, T_{max} is directly proportional to the
ratio of the beam power divided by the beam ratio P/r, and not
the power density, as was the case previously for homogeneous
beam profiles.

LAX has similarly shown that the temperature increase induced by a moving laser beam can be defined by [2.49]

$$T(t) = T_{max} \exp(vt/2r^2) \; I(vt/2r^2) \qquad (2.32)$$

where I( ) is the zeroth modified Bessel function. Further analyses of various different and related heating conditions can be obtained from the work of CARSLAW and JAEGER [2.48].

The theory presented so far is satisfactory in instances where the optical absorption is strong, i.e. in most metals at most laser wavelengths, and in semiconductors and insulators at wavelengths above their bandgaps.

Several groups have studied in detail the problem of cw laser heating of various materials, particularly silicon [2.50-53] in recent years. Silicon does not satisfy the conditions necessary for the above analysis to be used, since many of its thermal and optical properties are nonlinearly dependent upon temperature.

Two limiting cases will be described here. The first concerns the introduction of a strongly nonlinear thermal conductivity K(T) into the problem. In many instances where absorption is strong such that the source term in the heat equations can be described by the surface function described earlier, the thermal conductivity is the largest nonlinearity in the heat equation. For example a rational function can be fitted to the experimental data for silicon [2.54] by using the empirical fitting of NISSIM et al. [2.50]

$$K(T) = C_p D(T) = 299/(T-99) \;\; [W \; cm^{-1} \; K^{-1}] \; . \qquad (2.33)$$

Similarly for GaAs, the relationship can be written [2.55]

$$K(T) = 91/(T-91) \;\; [W \; cm^{-1} \; K^{-1}] \; . \qquad (2.34)$$

This knowledge can be incorporated into the solutions of the heat equation outlined above by employing a Kirchhoff transformation [2.56] to effectively map the original temperature (conveniently called the "linear temperature") to a new "true" temperature increase, the so-called "nonlinear temperature". With a similar knowledge of K(T) for any material whose heating properties are closely related to the present situation this transformation can be applied.

The transformation is applied using the formula

$$T_L = T_{max} = \int_{T_0}^{T_{NL}} \frac{K(T')}{K(T_0)} \, dT' \qquad (2.35)$$

Using the form shown above, the relation between the nonlinear T_{NL} and linear ($T_L = T_{max}$) temperatures is then

$$T_{NL} = 99 + (T_0 - 99) \exp[T_L/(T_0 - 99)] . \qquad (2.36)$$

Several measurements have actually been performed [2.57,58] in order to test the accuracy of this equation for the case of silicon. It is possible that the mild dependence of reflectivity upon temperature R(T), the weak deviation from the delta function approximation for the absorption of visible radiation, or even the small changes in density and specific heat with temperature, $\rho(T)$, and $C_p(T)$ could affect the overall accuracy of this relatively simple formulation. Strictly, of course, these second-order effects should properly be introduced into the problem from the beginning. However, the measurements showed that (2.36) could be accurately calibrated to the known melting point of e.g. silicon by inserting a constant N, into the exponential. Thus the actual temperature rise induced by argon laser radiation in single crystal silicon can be accurately estimated by [2.58].

$$T_{actual} = 99 + (T - 99) \exp[NT_L/(T_0 - 99)] \qquad (2.37)$$

where N = 0.96 was found to give the best fit. This formula is

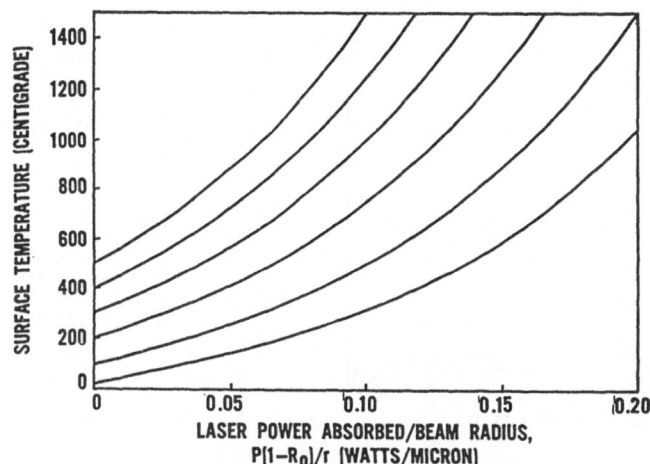

Fig.2.18. Temperature rise induced in silicon as a function of the ratio of absorbed argon laser beam power over the beam radius, for a range of initial substrate preheating temperatures using (2.37) in the text [2.95]

plotted in Fig.2.18 where the temperature of the sample is shown as a function of the ratio $P(1-R)/r$, given in units of $W/\mu m$, for a range of pre-chosen substrate temperatures.

Such a calibration has not yet been performed for GaAs, although it is expected that the calibration factor will be similarly close to unity. Nevertheless, a plot of induced temperature on the surface of GaAs under the same irradiation conditions is shown in Fig.2.19.

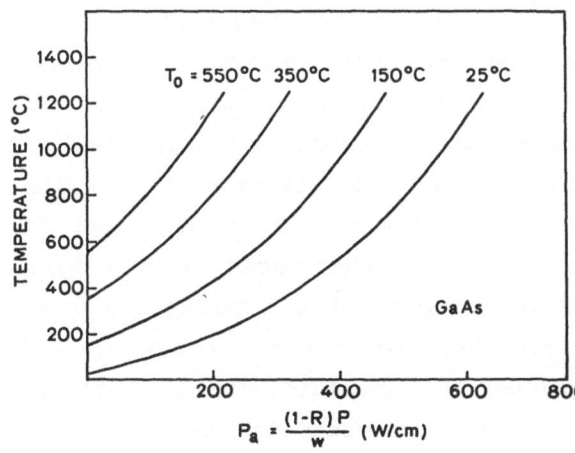

Fig.2.19. Temperature rise induced in gallium arsenide as a function of absorbed power per unit radius from the argon laser [2.87]

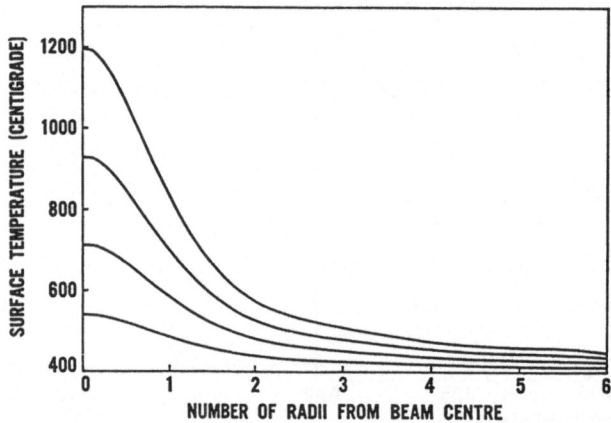

<u>Fig.2.20</u>. Spatial profile of the temperature gradients induced by argon laser on silicon as a function of beam radius for various peak temperatures [2.96]

The spatial profile of the laser induced temperature is shown in Fig.2.20 for the case of silicon described above. It can be seen that the spread of the induced temperature profile (defined in this instance by the distance traversed across the sample before the temperature falls to 1/e of its peak value), is approximately 50 larger than the 1/e diameter defined by the intensity profile of the laser beam itself.

Finally it is useful to consider the special case where the incident radiation is not strongly absorbed in the workpiece at the beginning of the irradiation event. Such situations are usually complicated by the behaviour of the absorption in real time during the irradiation and subsequent heating of the material. In the vast majority of instances, the absorption of the radiation actually increases as temperature increases, or indeed, if pulsed radiation is used, as the intensity of the pulse increases with time towards the middle of the pulse. It must be remembered, however, that in other cases, such as when thin layers are manufactured on the substrate during the irradiation, these can sometimes decrease the reflection and of course simultaneously increase the absolute energy absorbed. As a typical example of the former case consider the absorption of CO_2-laser radiation in silicon. CO_2-laser radiation is only weakly absorbed in weakly doped c-Si at room temperature. The photon energy ($E = hc/\lambda$) at approximately 10 μm being 117 meV, is insufficient to promote carriers from the valence to the conduction band, and therefore the measured absorption coefficient of 2 cm^{-1} for this radiation [2.59] is strongly influenced by the number of free carriers introduced by any means into the conduction band. This implies that any photo- or thermally-generated carriers during irradiation will create a strong positive feedback to the coupling system and further increase the absorption. Consequently, although initial absorption may be extremely small, if laser radiation is provided in short enough pulses, the possibility of thermal runaway exists which will lead to uncontrollable absorption, heating, and possibly eventual melting of the sample, which in many applications may be completely undesirable.

It is useful to estimate the minimum time required for the irradiated sample to reach an equilibrium temperature at the surface. We have already calculated the temperature increase expected in the situation of $L_d \ll \alpha^{-1}$. To this derivation, the temperature dependence of the absorption coefficient $\alpha(T)$ can be introduced. For example, it has been found that up to some critical temperature [2.59]

$$\alpha(T) = a \exp(T/T_1) = 2 \cdot 10^{-2} \exp(T/110) \quad [cm^{-1}] \tag{2.38}$$

by an empirical fitting to recent absorption data for silicon at this wavelength. The solution is of the form

$$T(0,t) = T_0 + T_1 \ln\left[\frac{\alpha(T_0)P(1-R)t}{c_p \rho T_1 \pi r^2}\right]^{-1} \tag{2.39}$$

and this represents the fastest increase in temperature possible under this irradiation condition. For example, once the absorption length is reduced to below that of the thermal dif-

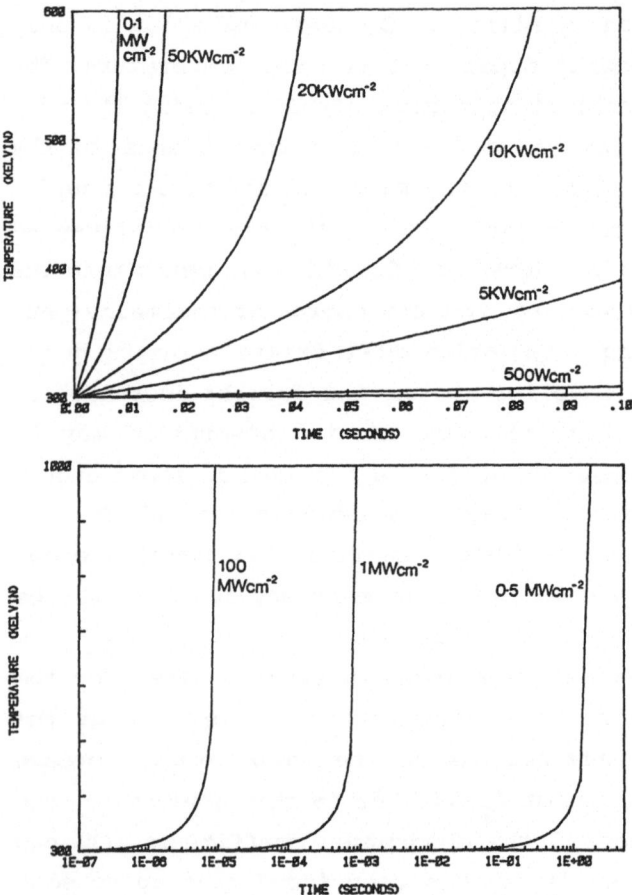

Fig.2.21. (a,b) Intensity dependence of the maximum rise in temperature induced in silicon by a CW carbon dioxide laser beam, shown as a function of time

fusion length, diffusion will become more important, and its influence will actually reduce the rate of heating. Nonetheless, as mentioned above, if the laser pulses are sufficiently short, diffusion will not have enough time to affect the peak temperature. In Fig.2.21, the intensity dependence of the heating cycle is shown, and it is clear that for typical pulsed laser intensities the heating is most rapid, and once initiated can accelerate and eventually lead to thermal runaway. For irradiance levels more typical of CW laser beams, the maximum heating rate is significantly slower, and the minimum process-

76

ing times for high temperature applications (i.e., up to 1300 K) are (at the very least) of the order of several hundred milli-seconds to several seconds. This is clearly a factor which should be taken into account when considering direct writing techniques using CO_2 lasers on materials that are at least in-itially weak absorbers of 10 μm radiation; it puts an upper limit of a few millimetres per second on the actual speed of beam scanning that can be achieved whilst still attaining the necessary temperatures.

So far in this section, we have chosen laser induced heating situations that have been described more or less straightfor-wardly by the usual heat equation, and have even discussed par-ticular circumstances where some simplifying assumptions can lead to useful special cases. Generally it is true that the less intense, and the longer the incident pulses are, the less nonlinear and more easily controllable the heating process becomes. This is why continuous beam processing has attracted much more attention than has the short pulse regime in surface chemical processing. It is interesting to note, however, that the shorter pulses can often introduce new and otherwise less accessible processing conditions which have often resulted in the discovery of novel and original phenomena that could ulti-mately be exploited.

In terms of modelling, if the deposition of energy from the laser beam induces a phase change in the workpiece, and if the deposition of energy continues after e.g. melting, or evapora-tion has occurred, the heating models must include the various latent heats of reaction, as well as the altered optical and thermal properties of the system. There are a large number of so-called "melting models" in the literature, only a fraction of which are referenced here [2.60-67]. However, when signifi-cant evaporation occurs near the boiling temperature of the material, modelling becomes appreciably more difficult. The energy that would otherwise be incident on the sample surface will now be scattered away from it by the evaporant rather

77

than being coupled into it. Therefore the nature of the evaporant, as well as any interaction with the environment, must be included in any model of the system. Although clearly not yet fully developed to its full potential, the controlled use of laser radiation to induce evaporation of atoms for film growth could grow significantly over the next few years. In this, and any other laser application, however, it would be useful to initially predict what irradiation conditions will be required for the operation.

Because of the large nonlinearities usually associated with laser heating of solids, we have already seen that it can be extremely difficult to accurately estimate the fluence required to achieve a required processing temperature for particular set of circumstances. Nevertheless, it is quite straightforward to ascertain when the melting point has been reached, by simple microscopic inspection of the surface after irradiation under carefully chosen processing conditions. Once this temperature, and fluence threshold have been determined, it usually becomes a relatively simple problem to calculate the additional fluence necessary to induce evaporation. This is a direct consequence of the fact that the optical and thermal properties of liquids, and particularly metallic liquids, are not so dramatically nonlinear as their solid-state counterparts. As previously discussed, the flow of heat in solids is primarily due to phonons and electrons, which can transport energy from one area to another, driven by density gradients, thermal profiles etc. However, when the number of free carriers in the material is appreciable, as in a highly excited semiconductor, a liquid or in a metal, the electronic contribution to the thermal conductivity is significant. This is known as the Wiedemann-Franz Law, written as

$$K_e = k_0 \dot{k}^2 \sigma T/e^2 \tag{2.40}$$

where K_e is the thermal conductivity due to the electrons, σ is the electrical conductivity, k_0 is a constant, T is temperature, k is Boltzmann's constant, and e is the electronic charge. Therefore, when σ is proportional to T_D/T, i.e. at temperatures greater than the Debye temperature, T_D, the thermal conductivity can be considered to be essentially constant with temperature. In insulating materials, heat is transported primarily by the diffusion of phonons from the high to the lower temperature regions of the material. These phonon-phonon scattering processes are extremely complicated, but a very much simplified analysis yields the following relationship

$$K_p \simeq \exp[\hbar\omega(q)/kT] - 1 . \tag{2.41}$$

Therefore at high temperatures, we can predict that the thermal conductivity is inversely proportional to the temperature, when phonon-phonon scattering dominates.

When there are no losses of energy during the heating, i.e. when ultrashort pulses are used, or alternatively when thermal conductivity is negligible, the incident fluence F required to evaporate one absorption depth of material once melting has occurred can be written

$$F = [C_p\rho(T_b - T_m) + L]\cdot(1 - R)/\alpha \tag{2.42}$$

where L is the latent heat of evaporation, and $(T_b - T_m)$ is the temperature difference between the melting and boiling (evaporation) points.

This situation is best shown for silicon irradiated by picosecond and subpicosecond pulses at various wavelengths (Table 2.3). It can be seen, for example, that although a spread in incident fluences of nearly 2 orders of magnitude between irradiating with UV and near IR wavelength radiation is necessary to induce melting, almost the same additional energy is required by all wavelengths to bring about noticeable evaporation in each case [2.68].

When longer pulses of radiation are applied, the situation reverts to that shown previously in (2.24) at the beginning of

Table 2.3 Data for ultrashort pulse melting and evaporation of silicon

Wavelength nm	Pulse width ps	Melting threshold J/cm^2	Additional fluence to evaporate melted material J/cm^2	Reference
1060	48	1.7	0.1	Boyd et al. [2.56]
620	0.09	0.1	0.15	Downer et al. [2.104]
533	20	0.2	0.2	Liu et al. [2.105]
248	15	0.024	0.12	Bokor & Bucksbaum [2.106]

this section, when absorption is confined to the surface layers and the temperature is controlled by diffusion away from the surface into the bulk. A rearrangement of this equation enables one to calculate the time needed to reach the evaporation point t_b for various materials. This is given by [2.69,70]

$$t_b = (\pi/4) \cdot (K\rho C_p/F^2) \cdot (T_b - T_m) \tag{2.43}$$

where F is the absorbed power density, $F = P(1-R)/\pi r^2$. As discussed above, this situation is most relevant to metals and metallic-type media. It is not surprising, therefore, that such formulae are useful in applications where evaporation of metals is performed, e.g., in cutting and drilling. Obviously this formulation can also be applied to other materials that do not appreciably change their physical properties between any two temperatures of interest, such that (T_b-T_m) could be redefined to be some other incremental change in temperature.

The amount of evaporated material does not increase monotonically with increasing peak power in the laser beam. Once evaporation occurs, the ejected material may absorb any remaining energy impinging on to the material, and can heat up dramatically. The material subsequently becomes slightly thermally ionized, and is thereafter shielded from the radiation by this high temperature plasma. As discussed in Chap.6 this situation leads to a new regime of laser processing of thin films, namely, *Laser Pulsed Plasma Chemistry* (LPPC).

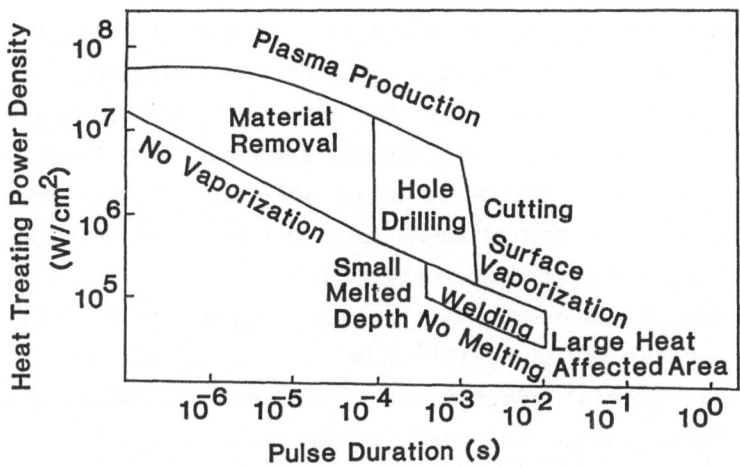

Fig.2.22. Regimes of suitable laser beam intensities and pulse durations for a range of established laser applications [2.70]

Figure 2.22 shows the well-known menu of laser processing regimes in terms of incident power density and pulse duration [2.70]. It is important to note that this is by no means a universal mapping of the various regimes, but more accurately an indication of the relationship between the requirements of each process. Obviously, the regions will vary considerably between materials, environments, wavelengths, pulse profiles etc. Having discussed heat treating in the solid phase, a process usually handled by cw lasers, and evaporation at the other extreme, inevitably a short pulse laser application, we will now discuss the middle region of laser induced melting. This regime, although accessible with both types of lasers, is usually a domain of the pulsed laser.

The strong variations of C_p, K, R, ρ, α, and D with temperature, time, depth into the material, and phase, during pulse laser melting of materials, mean that a simple analytical solution of the heat equation is not possible, even for a general case. This has resulted in the publication over the past few years of a large number of theoretical models involving numerical solutions of particular heating and melting situations, as referenced earlier [2.60-67]. The publication of CARSLAW and JAEGER

81

is also a valuable source of reference in this area [2.48]. It is interesting, nevertheless, to note some of the numerical information produced by these models. For example, with nanosecond melting of silicon using 10 ns pulses from the ruby laser, heating and cooling rates of the order of 10^{11} and 10^9 K/s are predicted, while melting can proceed into the material with a planar front velocity of approximately 10 m/s.

For picosecond and sub-picosecond pulses inducing melting in silicon, heating rates of up to 10^{15} K/s are achievable, sometimes leading to material ejection on a picosecond timescale, and resolidification rates that are so fast that ordered crystallization cannot occur. This phenomenon of laser-induced amorphization has also been shown to occur with nanosecond pulses at wavelengths which are strongly absorbed at the surface of the sample [2.69].

One of the main differences between laser beam induced heating of semiconductors and metals, is the rapid cooling times available with metals, arising from their higher thermal conductivity. As has been pointed out above, the heating and cooling rates of silicon can be different by about 2 orders of magnitude. For aluminium on the other hand, these figures have been estimated to be $4 \cdot 10^{10}$, and 10^{10} K/s, respectively.

Since pulsed lasers allow precise control of the melt duration, this enables accurate relocation of implanted impurities. Also dopant can be driven into the substrate either from the gas phase, after some sort of laser CVD, for example, or from a layer painted onto the surface. Sometimes two laser beams can be used in this procedure, one to prepare the species, and the other to drive it into the material, and this gives rise to the field of laser doping.

2.5.3 Impurity Incorporation

As a consequence of the rapid heating, melting and cooling mentioned above, regrowth velocities of more than 10 m/s can be achieved using lasers. In many cases, crystalline material can be reformed during this extremely rapid process, compared with crystal growth velocities of around several mm/hour that are more typical of conventional crystal production. Under the laser irradiation conditions, a monolayer of atoms must be grown on to the recrystallizing interface in times of the order of 10^{-10} s. Nevertheless, it has been estimated that whilst in the liquid phase, the atoms of the melt can impinge upon the solid surface in times around 10^{-12} s. It is in fact possible to cool the melt so fast that the atoms do not have sufficient time to find the position of least energy within the growing interface, and non-ideal crystalline structure or even an amorphous structure can be formed as discussed above. If impurities are present within the gaseous environment during this process, there will be a possibility that they become incorporated in the regrown material. Likewise, if atoms are already present within the workpiece, they too may become embedded in the newly formed structure.

Such impurities, however, are subjected to various other effects before the material actually resolidifies. They may diffuse, or segregate, for example. The extent to which diffusion occurs will depend upon, amongst other factors, the velocity of solidification, the solid solubility limit, and the segregation coefficient k defined as C_S/C_L, where C_S and C_L are the impurity concentrations in the solid and liquid phases, respectively, at the interface. The incorporation of impurities in rapidly melted and solidified material is both very efficient and can, in fact, create such a density of substitutionally incorporated species that the normal solid solubility limit can be exceeded, in some cases by several orders of magnitude. Such phases, having been formed under non-equilibrium conditions, are generally metastable in nature, and are often called solid solutions.

It is attractive to consider the formation of these structures through the impurities not being given sufficient time to diffuse before being trapped behind the rapidly advancing resolidification front. However, it is important to remember that this is only a convenient and strictly quite an oversimplified view of the situation. One must additionally consider the influence of the melt-front velocity, the degree of undercooling, and the sticking coefficient of the different species of atoms at the interface to the substrate. Under the effect of these non-equilibrium phenomena, the use of k is no longer strictly valid, and one uses the parameter k', extracted from experimental data.

If the value of k' is low (e.g., < 1) for a particular species in some material, then by implanting concentrations above the solid solubility and annealing by short laser pulses the solid solubility of the atoms in that material could be exceeded. If k' is very low (e.g., 10^{-4}), then it is possible to create a zone refined region in which impurities have been rejected by the resolidifying material in such a way that an extreme accumulation of rejected atoms builds up at the surface of the material.

Under certain conditions, a local nonuniformity in the distribution of impurities can lead to a local disturbance of the thermodynamic properties of the material such that observable perturbations in the resolidification occur. For example, the phenomenon of constitutional supercooling [2.71] arises directly from an instability in the resolidification front caused by a nonuniform distribution of an impurity that lowers the freezing temperature, at some local point in the material. While this region remains molten slightly longer than the rest of the sample, it must accept the excess atoms diffusing from the solidifying surroundings, and this further delays solidification at that point, since the freezing level is also further reduced. When the concentration of impurity exceeds the maximum that can be incorporated in to substitutional sites in the solid, the excess forms a network of interlinked cells that defines an undesirable periodic structure of separated phases. The implantation of nitrogen in silicon has been found to show

84

this effect, as will be discussed later in Chap.5, as has Ga, Fe, In, Cu and Pt, to name but a few [2.71].

2.6 Nucleation and Growth

Three principal mechanisms of thin film formation have been identified over the years. The first, the van der Merwe mechanism, is characterized by a layer-by-layer growth; the second, is a three dimensional nucleation, growth and coalescence of islands (the Volmer-Weber mechanism) and the third, operating via the adsorption of a monolayer and subsequent nucleation on top of this layer, is known as the Stranski-Krastanov mechanism [2.72]. In the majority of cases, the second listed process occurs, and this is the one thought to be closest to that operating in many of the laser induced deposition techniques.

The individual steps comprising the growth of a thin film from the gas phase include impingement and sticking of the components on to the surface, surface migration of the adsorbed species, nucleation of the species, clustering of individual nuclei into interconnecting islands, and then filling in of the network spacings, before uniform layers can be formed. Some films may be 10's to 100's of nm thick before this final stage may be considered near to completion. Nucleation can often be influenced by the presence of various surface structures, such as point defects, or crystal dislocations and other imperfections; in fact this property is often used to decorate such features on crystal surfaces in order to render them visible. Clearly, the initial stages of this film-forming routine are crucial to the eventual outcome of the film structure and quality, and the presence of specific photons during the process could be beneficial or disastrous to either. Having discussed the various ways in which particles may be adsorbed onto (and desorbed from) a solid surface in the previous section, we will now turn our attention to nucleation phenomena.

Nucleation and growth has often been observed by electron microscopy for metal vapours condensing on to insulating sub-

strates at room temperature. By all accounts, the initial seeding process must be random in nature. Yet most theoretical
treatments [2.73,74] are based on deterministic rate equations
first derived by ZINSMEISTER [2.75], which characteristically
yield a unique solution for each set of initial conditions. Particles will naturally impinge upon a surface from a vapour in a
random fashion, and assuming that the sticking coefficient s is
large. (In practice, $s \simeq 1$ if the desorption energy of the particle on the substrate is <4 of its incident energy [2.72]. If a
particular species is attracted by the fields at the surface, it
may remain there for some time, able to move around the surface until either losing excess energy to a defect, thereby
becoming temporarily bound, or gaining sufficient energy to
escape. The time spent on any weakly bonded sites, t_b, can be
written

$$t_b = t_0 \exp(\phi_m/kT) \tag{2.44}$$

where ϕ_m is the activation energy for migration, and t_0 is the
inverse surface vibrational frequency of the adatoms. The total
time spent on the surface is determined by the desorption activation energy ϕ_d and is of the form

$$t = t_0 \exp(\phi_d/kT) \tag{2.45}$$

The value of t_b is largest at low sample temperatures, where
the particles easily lose all their excess energy very quickly,
and become bonded. At increased temperatures, the species may
occasionally move from site to site in a random fashion, until
at very high temperatures, t_b is so small that the adsorbed
particles act as a two dimensional gas on the substrate.
Clearly, nucleation can only occur if there are sites at which
the particles can lose their excess energy. The precise nature
of these preferred sites still remains a relatively unexplored
regime of thin film preparation, and even more so in laser
assisted growth.

86

The re-evaporation of desorbed species will also obviously affect the nucleation rate. If the overall rate of adsorption exceeds the desorption rate, then the surface will soon become supersaturated, and nucleation will become more probable. However, the controlling factor is the ratio of ϕ_d/ϕ_s, where ϕ_s is the heat of sublimation of the condensed film. For small ratios, nucleation occurs without supersaturation and the coverage can be quite high, whereas if $\phi_d > \phi_s$, then condensation will only occur if a high degree of supersaturation is achieved. At a value around 1, the classical theory of nucleation, based on thermodynamic concepts (capillary theory) can be satisfactorily applied [2.72].

There will inevitably be important interplay between nucleation rate and surface migration. For example, without surface diffusion, it is easy to perceive a much slower cluster formation rate, determined only by direct impingement of particles on to the surface. On the other hand, surface migration should accelerate the cluster formation and subsequent film growth. In fact, the actual cluster density may be largely determined by the migration processes. After a certain density of clusters has been formed, either directly, or by migration, subsequent diffusion of species may serve a dual purpose of not only finding new nucleation centres, but also supplementing the growth of existing clusters and assisting in the coalescing of neighbouring clusters.

The growth of clusters obviously relies upon the formation of bonds when an impinging or migrating atom encounters one of these new centres, or a cluster. In fact, aggregate disintegration may also occur simultaneously. The cluster stability will depend upon many parameters, such as surface tension forces at the substrate, the bonding relationship between aggregate atoms and substrate, and the energy difference between bonded and unbonded species, as well as ambient pressure and temperature.

Existing clusters, acting as established seeding sites for

incident particles, will continue to grow, in three dimensions, eventually forming large "islands". These individual islands will also coalesce when their boundaries begin to overlap. It is thought that in many cases sufficient heat may be liberated upon coalescence so as to melt the islands (being microscopic clusters, their melting point could be significantly smaller than that of the bulk film). Clearly, the eventual quality and structure of the grown film will be dictated not only by such processes, but by the relationship between the various thermal properties of the film and substrate.

The eventual nucleation rate J, determined by the classical capillary theory, will depend upon N^*, the concentration of aggregates over a critical size, given by

$$N^* = n_0 \exp(-\delta G^*/kT) \qquad (2.46)$$

where δG^* is the corresponding critical energy of the nucleus formation and n_0 is the density of adsorption sites. It will also depend on Γ, the rate at which molecules join the critical nucleus by surface diffusion. (Those joining by direct impingement are negligible at this stage). Thus, the nucleation rate can be written

$$J = 2\pi r^* Z \Gamma N^* \sin\theta \qquad (2.47)$$

where r^* is the radius of the critical nucleus, θ is the contact angle of the aggregate with the substrate, and Z is Zeldovich's constant ($\simeq 0.01$). Whilst this theory provides a qualitative view of the main mechanisms of thin film formation, it uses thermodynamic concepts that more correctly apply to macroscopic systems. Since many more recent studies are involved with sub-nanometre dimensions, more specialized alternative theories are clearly required.

2.7 Chemical Reactions and Growth Rates

The range of mechanisms by which the laser can initiate chemical reactions has already been addressed in this chapter. However the preparation of the substrate or the gas phase species for reaction is only one step in a chain of events that must be rigorously followed before the desired chemical interactions occur at all. Once it has been found that the required chemical reaction can be initiated, the rate at which it proceeds at any future time can be modified by altering, amongst other things:

1) The irradiation geometry.
2) The properties of the laser radiation, e.g. wavelength, pulse width, intensity, spot size etc.
3) The composition of the gaseous environment.
4) The properties of the gaseous mixture, e.g. pressure, temperature, flow rate etc.
5) The properties of the substrate, e.g. temperature, surface structure, preparation etc.
6) Transport of the species, which may be affected by diffusion, molecular flow, convection etc.

It is not difficult to see that the effect of altering some of these processing conditions by too much from the norm could actually provide a means of altering the stoichiometry of the structures formed, and also the precise composition of the layers themselves. Indeed it is possible that the laser can be used to form exotic multilayered structures with very fine control over the thickness of each layer.

Figures 2.23,24 give a general breakdown of the simple step by step processes that could be involved in the formation of thin films and microstructures by laser. One could consider, for example, laser induced etching, deposition, growth, alloying etc. that could be described by either of the flow charts, although it is clear that not every step is essential for every operation. To calculate the reaction rate from a purely theoretical basis would be a monumental task for some processes. However, if the one or two limiting steps in the operation could be identified, a very good approximation for the expected reaction rate could usually be expected.

89

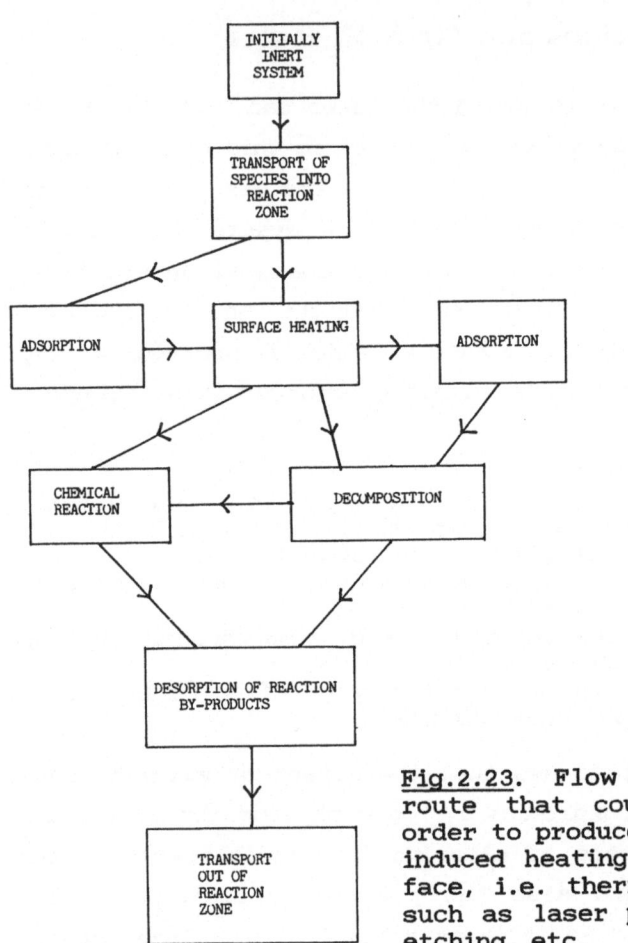

Fig.2.23. Flow diagram of a general route that could be followed in order to produce thin films by laser induced heating of the sample surface, i.e. thermochemical reactions such as laser pyrolysis, oxidation, etching, etc.

2.7.1 Low Intensity Levels

At low levels of incident laser beam intensity, one can usually neglect any depletion effects on the source of the parent molecule. The production of reactant can then be related to the average intensity of the laser beam, I, the density of parent molecules, N_p, if the reaction is photolytic, the photodissociation cross-section at the wavelength used, and the radius of the laser beam at focus as long as this is much less than the depth of focus. A second fairly reasonable and general assumption is that at these low levels of irradiance, the reaction is

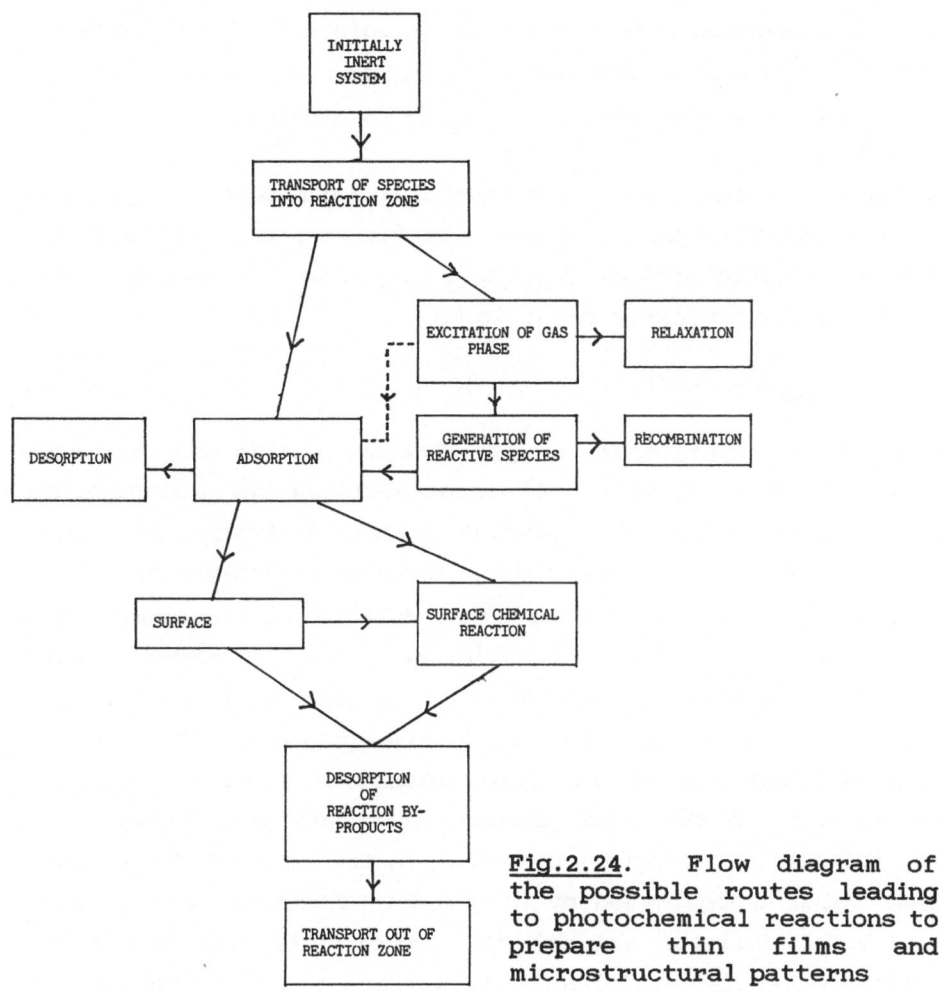

Fig.2.24. Flow diagram of the possible routes leading to photochemical reactions to prepare thin films and microstructural patterns

not limited by diffusion but by the rate of chemical adsorption, desorption, or reaction at the sample surface. This assumption is independent of whether the reaction is pyrolytic or photolytic. EHRLICH and TSAO [2.76] using precisely these assumptions, have shown that at the centre of the laser beam on the sample surface, the reaction rate due to gas phase processes, $j_{gas\ phase}$, is approximated by

$$j_{gas\ phase} \simeq k_g I N_p(P) \sigma w_0 \beta/2 \,, \tag{2.48}$$

91

i.e., the reaction rate for gas phase molecules is proportional to w_0^{-1}, (through Iw_0) where w_0 is defined as the beam waist, β the sticking probability, and k_g is a geometrically-related constant.

For processes in which the controlling photochemistry occurs in an adsorbed layer on the sample surface, the reaction rate for the adsorbed phase, $j_{adsorbed\ phase}$ at the centre of the laser beam is similarly shown to be

$$j_{ads\ phase} \simeq I\sigma\theta(P,T) , \qquad\qquad (2.49)$$

i.e., it is proportional to w_0^{-2}. The assumption has been made here that the photochemical cross-sections per molecule are independent of surface coverage $\theta(P,T)$. Although not always strictly true, this is a useful simplifying assumption.

The dependence of the reaction rate in the adsorbed phase being proportional to the beam intensity (power/radius squared) is easily understood. The reaction in the gas phase having a dependence on power/radius is a consequence of the reactant species being created in three dimensions above the surface, rather than in the "two" dimensional monolayer. Indeed, TSAO and EHRLICH [2.76] have shown that in the case of the incident beam being approximated by a cylinder of radius w_0 and essentially of infinite extent, the geometrical distribution of excited molecules arriving at the surface can be written as

$$k_g = (\text{solid angle})/4\pi = (1 - \frac{2}{\pi})\tan^{-1}(\frac{r/2w_0}{8}) . \qquad (2.50)$$

We see that at the beam centre (r=0) this tends towards $k_g = 1/8$.

Further study of the equations reveals that for gas phase reactions, the rate is linearly dependent upon the density of the parent molecules, which in turn is directly proportional to the pressure. Therefore a linear increase in pressure will result in a corresponding linear increase in reaction rate.

92

Whilst the reaction is controlled by this mechanism, and (2.50) remains valid, there should not be any strong dependence on the temperature.

By contrast, reactions dominated by adsorption mechanisms are quite strongly dependent on the degree of coverage of the adsorbent, and this in turn is significantly influenced by both temperature and pressure. As already pointed out in Sect.2.4, except for very low pressures and monolayer adsorption or chemisorption, the coverage of molecules on to the surface is very often a nonlinear function of pressure.

Adsorption, which by its nature involves some degree of bonding, results in a reduction in the energy of the adsorbate-surface complex energy, and therefore is by all accounts an exothermic reaction. By analogy, the reverse reaction of desorption is an endothermic reaction, requiring energy to function, as previously discussed. Therefore, a reduction in temperature will favour the forward reaction of adsorption. It is indeed well known that one of the easier methods of increasing the amount of adsorption of a surface is to decrease its temperature.

2.7.2 High Intensity Levels

An attraction of laser-induced film formation and microstructural definition is the extremely fast reaction rates that arise from the previously unattainable levels of irradiance which are presently available from laser systems. In fact, a potential drawback in many of the gas phase and surface phase reactions initiated by laser beams is the rapid local depletion of donor molecules that supply the reactive species for the reactions. In this regime, therefore, it is not the actual rate of the microchemical reactions that limit the complete process, but the availability of reactive species in the required area of processing. In other words, the mass transport into the reaction zone will become the rate limiting step.

Having identified the ultimate rate–limiting factor of these laser–controlled processes, it must be mentioned that in some notorious cases there is not yet the opportunity of reaching such reaction speeds that depletion of the supply of reactant becomes a problem. For example, in photodeposition, where the wavelength of the laser radiation is well matched to the absorption spectra of the donor molecules, the deposition rates are restricted by the gas phase nucleation processes that produce copious quantities of undesirable powdery deposits. The most usual way of avoiding such effects involves the introduction of an inert buffer gas such as N_2, He, or Ar. This, however, also indirectly reduces the deposition rate of the more desirable film structure. In a great many other cases, the highest reaction rates achievable are not even close to those that incur problems such as that described above. This is usually a consequence of the poor wavelength match between precursor and laser beam. This is obviously a dual problem of the unavailability of laser wavelengths at the appropriate absorption band of "off the shelf" precursor molecules and also the lack of appropriate molecules whose structural make–up can efficiently absorb the few and discrete laser wavelengths presently commercially available (Chap.3).

As a consequence of these circumstances, the regime of mass transport limited gas phase microreactions is most likely going to remain quite inaccessible for laser photolysis whilst the quality and degree of control over the process remain the most desirable operating requirements. Nevertheless, typical photodeposition systems free of uncontrolled nucleation in the gas phase can present growth rates in the region of 2 μm/s.

In situations where gas phase effects are relatively unimportant, and the surface reaction is significant, the reaction will usually be limited by mass transport. As EHRLICH and TSAO have shown [2.43], if the laser induced reaction is limited to a small area on the surface, the mass transport must be modelled by a full three dimensional diffusion equation, since the reaction now enjoys a supply of reactant from a small zone extending radially outward from its centre. Conventional large area planar reactions, by comparison, that are diffusion controlled by supply of reactant or release of products, can be modelled by using only the one dimensional diffusion equation. By considering only gradients due to concentration (and not temperature) in the external medium), leading to a constant diffusivity D and a reaction rate which is proportional only to the surface concentration of reactants within a small hemispherical zone of radius w_0, the steady state surface reaction rate j_{surf} is given by [2.43]

$$j_{surf}(t \rightarrow \infty) = \frac{Dn_\infty}{r_0 + w_0} \tag{2.51}$$

where $r_0 = 2D/fv$ is related to the mean free path of molecules in the gas, v is the rms velocity of the gas molecules, f is the fraction of collisions of the surface that result in a reaction, and n_∞ is the fixed molecular density away from the film. This equation is plotted in Fig.2.25 as a function of pressure for

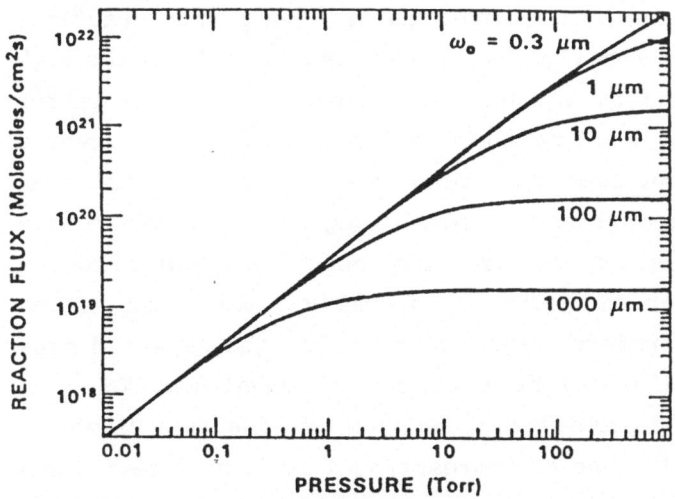

Fig.2.25. Computed steady-state values for the molecular reaction flux as a function of pressure given by (2.51) for various values of spotsize. The curves are typical of thermally activated laser etching. (See [2.43] for more details)

various values of w_0. Under most conditions the steady state conditions were reached within a millisecond for $w_0 = 100$ μm, within a microsecond for $w_0 = 10$ μm, and significantly more rapidly as beam dimensions approached 1 μm. In fact, the steady-state time t_{ss} can be written as

$$t_{ss} \simeq \frac{1}{D} \left(\frac{1}{r_0} + \frac{1}{w_0} \right)^{-2} . \tag{2.52}$$

For small spots, $r_0 \gg w_0$, the reaction rate j_{surf} (t→∞) is given by

$$j_{surf}(t \to \infty) \to Dn_\infty / r_0 = fvn_\infty / 2 \tag{2.53}$$

and is limited by interfacial kinetics, is almost linear with pressure, and is independent of w_0. For the other extreme, where $r_0 \ll w_0$, the surface reaction rate can be written

$$j_{(surf)(t \to \infty)} \to Dn_\infty / w_0 . \tag{2.54}$$

Here it is limited by diffusion kinetics, but is independent of pressure since D is proportional to $1/P$, and is inversely proportional to w_0. A bold comparison between planar reactions and those initiated locally by a laser beam would, in the high pressure limit, lead to a predicted enhancement of the latter over the former by factors approaching infinity! Of course, a more realistic comparison must take into account the fact that most planar film growth occurs in flowing reactors, where diffusion of the necessary species only needs to occur across a thin boundary layer of width δ, typically 3 mm in an atmospheric reactor. Therefore, using $w_0 \simeq 0.3$ μm, the expected geometric enhancement would be 4 orders of magnitude. In fact, growth rates during pyrolytic deposition of some thin films, as will be discussed in Chap.6, approaching 1 mm/s have been achievable by taking advantage of this possibility.

Just as gas phase diffusion to the surface can be one mass transport limiting case, surface phase diffusion can similarly hold up the laser induced reaction. TSAO and EHRLICH [2.76] once more have published details on this regime. Basically, at sufficiently high laser intensities, the adsorbed molecules are rapidly transformed into their relevant products, and before any more reaction can occur, more reactant must diffuse into the zone influenced by the laser spot. At lower intensities it is believed that such diffusion will not be noticable, and the reaction will be photon-limited. The authors suggest that by monitoring the reaction rate under these conditions, the surface diffusivities of various molecules could be determined.

2.7.3 CW Laser Controlled Reaction Rates

Many thin film reactions can be characterized by one of two types of growth processes: by interface or bond-breaking kinetics which tend to represent linear growth rates, or by a diffusion controlled reaction, where the transport of reactant to the reaction site usually exhibits parabolic-type behaviour. Diffusion limited reactions can be fitted by [2.77]

$$d^2(t) = \beta t \, \exp(-E_a/kT) \qquad (2.55)$$

while film formation controlled by a rate limited processes can be represented as

$$d(t) = Bt \, \exp(-E_a/kT) \qquad (2.56)$$

where $d(t)$ is the reacted thickness at time t, B and β, are constants, E_a is the activation energy, and k and T have their usual meanings.

The growth of certain silicides (e.g.,Ni,Pt,Pd) is governed mainly by the diffusion equation of (2.55) above, while the regrowth of a crystal being annealed can be described by (2.56) above. The oxidation of silicon, on the other hand, as we will see later, is subtle mixture of both linear and parabolic rates, but for the growth of sufficiently thin layers (films of thickness at least up to 20 nm) the linear growth rate provides an excellent approximation.

For a stationary laser beam, providing growth is not appreciable during the time taken for the substrate to come into thermal equilibrium with the incident heat source, the application of the above equations provide no significant problems. However, if a laser beam is scanned over a larger area such that any point on the workpiece is subjected to a series of rise and fall temperature fluctuations as the laser beam approaches and leaves the point, then (2.56) is more accurately written

$$d(t) = B \int_0^t \exp[-E_a/kT(t')] \, dt' \qquad (2.57)$$

where $T(t')$ is the time varying temperature induced at the point of interest by the moving laser beam. This parameter may be conveniently defined by an effective temperature $T_{eff} = T$, already calculated in the previous section, and an effective time $t_{eff} = f_r t_{dwell}$, where t_{dwell} is the dwell time of the laser beam at any point, and f_r is the "dwell time reduction factor" defined mathematically elsewhere [2.78-80]. For example, f_r was calculated to be approximately 0.45 for the case of laser induced oxidation of silicon, which will be discussed later in Chap.4. Equation (2.57) above will also apply for cases where a short pulse of laser radiation is used, and the sample is similarly subjected to a cycle of temperature increase and decrease.

For the case of multiple overlapping scans, the reacted layer produced at a given time is a summation of not only all the scans passing directly through that point, but also many neighbouring scans that passed nearby and raised the temperature slightly. In this case the t_{eff} defined above is no longer valid. An alternative approach, applicable only where uniform films are produced, is to define the average reacted thickness <d> prepared in time t, divided by the total area A scanned by the laser beam. This can be written

$$\langle d \rangle = A^{-1} \int_0^\infty d(r) \; 2\pi r \; dr = \frac{2Bt}{A} \int_0^\infty \exp[-E_a/kT(r)] \; dr \qquad (2.58)$$

for reaction limited kinetics. Here, as before, the temperature can be equated to the maximum induced temperature, T, but the effective time t_{eff} is redefined as [2.77,78]

$$t_{eff} = 4tr^2 \; f_r^2/A \; . \qquad (2.59)$$

Similar derivations can be employed for other types of reactions governed by, for example, logarithmic and/or power laws as are found in the oxidation of metals.

Although quite preliminary, these calculations of the MIT group present a extremely useful overall picture of the various processing regimes now available with small volume laser induced chemical reactions. The vastly superior reaction rates can be readily explained, and extrapolated to define the ultimate processing limitations of the various preparation techniques. Under fine control, these film growth or removal methods could also be used to measure relevant physical properties, such as surface diffusion, of various species.

3. Experimental Considerations

In this chapter the essential experimental criteria for laser processing are considered. Firstly, the basic properties of laser beams are briefly reviewed, including the definitions of spot size, divergence, depth of focus etc, as well as the characteristics of beam propagation. Optical resolution is also discussed, not only in terms of the properties of the laser radiation, but also particular physical and chemical mechanisms that ultimately can control spatial line widths. Then the many varied modes of laser processing that have been used over the past few years are introduced and summarised. Particular attention is given here to pattern replication methods. The practical aspects of process uniformity and repeatability, as well as beam profile engineering and measurement are also described in this chapter.

The choice of lasers available for laser processing is included for completeness, to assist those not familiar with present commercially available systems. A complete summary of the most commonly used lasers and their relevant specifications is given, along with the comparative merits of each in various applications.

3.1 Properties of Laser Beams

A laser cavity can often support a number of transverse electromagnetic modes (TEM's), once the lasing threshold has been surmounted. If one does not discriminate, it will often operate

in a multimode fashion, with several TEM patterns being simultaneously supported within the cavity, leading to an arbitrary
summation of the modes making up the overall spatial profile of
the laser beam. However, it is usually relatively straightforward to discriminate against all but the fundamental mode, the
TEM_{00}, by placing an aperture inside the cavity, thereby allow-
ing only this smallest mode to propagate. This Gaussian profile
is by far the most desirable and widely encountered in lasers.
It most usually requires the least energy to support it, and it
has a smaller beam divergence than the higher order modes, and
is very symmetrical. Furthermore, as it propagates through various optical systems, it retains its near and far field Gaussian
form, unlike any of the higher modes.

The Gaussian beam has a peak energy content in the centre,
and as the subscripts in the TEM definition suggest, has no
null-points across the xy plane. The actual energy distribution
within the TEM_{00} beam can be written as

$$E(x,y) = E_0 \exp(-r^2/w_E^2) \qquad (3.1)$$

where E_0 is the peak field amplitude, $r^2 = x^2 = y^2$, and w_E is a
constant which, when $r = w_E$, defines the point where the value
$E(x,y) = E_0/e$ (Fig. 3.1). Alternatively, since intensity, or
irradiance, $I(x,y)$ is related to the square of the electric
field, the beam profile can be defined as

$$I(x,y) = I_0 \exp(-2r^2/w_E^2) \ . \qquad (3.2)$$

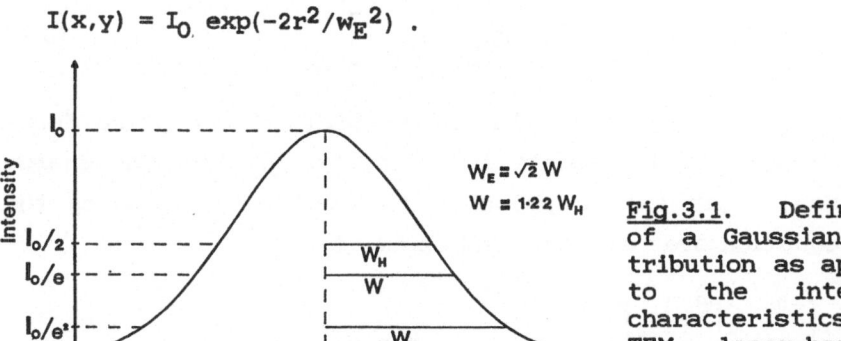

$W_E = \sqrt{2}\,W$
$W \equiv 1\cdot22\,W_H$

Fig.3.1. Definition
of a Gaussian dis-
tribution as applied
to the intensity
characteristics of a
TEM_{00} laser beam

Therefore, at the same value of r, the intensity has decreased to the value of I_0/e^2. The distance from the beam centre to either the $1/e$ of the field, or the $1/e^2$ of the intensity, defines what is commonly known as the spot radius. The spot size can be defined as $2w_E$. It is important to note that other definitions of beam dimension are sometimes used, such as where the ratio $I(x,y)/I_0 = 1/e$, or 0.5, etc. Such a variety of alternatives can often lead to confusion, and the reader is advised to tread warily amongst these variously related definitions of the same beam profile. The total power P within such a laser beam is given by

$$P = \pi w_E^2 I_0/2 \ . \tag{3.3}$$

Therefore, if one defines the irradiance distribution as

$$I(x,y) = I_0 \exp(-r^2/w^2) \tag{3.4}$$

then the total power may be written

$$P = \pi w^2 I_0 \ . \tag{3.5}$$

In this case the radius used, w, defines the $1/e$ point of the intensity, and $w = w_E/\sqrt{2}$. If one now considers a short pulse of radiation, whose temporal duration is also Gaussian in nature, i.e.

$$I(r,t) = I_0 \exp[-(r/w)^2] \exp\left[-(t/t_p)^2\right] \tag{3.6}$$

then t_p similarly defines the "half-width" of the pulse duration between the $1/e$ points in intensity, in time. The energy contained within such a pulse is obtained by integrating the above equation over all time and space, i.e.

$$E = \int_0^{2\pi} d\theta \int_0^\infty r dr \int_0^\infty I(r,t) \ dt = w^2 \pi^{3/2} t_p I_0 \ . \tag{3.7}$$

Additionally, the fluence F (sometimes referred to as energy density or flux) can be defined as $E/\pi w^2$.

In the introduction, several advantages of the application of Gaussian laser beams were discussed. These included the beam divergence, the focusing properties in terms of minimum spot size achievable and the depth of focus, coherence, wavelength selectivity, and bandwidth. The appreciable degree of coherence available in most lasers unfortunately cannot, or rather has not yet, been exploited in the area of thin film processing although it produces spot sizes smaller than otherwise identical but incoherent radiation can. The wavelength selectivity will be discussed separately in the next section, although the bandwidth, useful in the selectivity of particular reactions, is not an essential topic for discussion here.

The actual degree of laser-beam divergence can be controlled somewhat by certain design criteria on the oscillator cavity, but the fundamental limit to the propagation is naturally defined by the physical phenomenon of diffraction. For example, in the case of single-mode beams whose energy distribution in space is Gaussian, the angle of divergence is described by

$$\phi = 2\lambda/\pi d \tag{3.8}$$

where ϕ is given in radians, λ is the wavelength, and d is the diameter of the limiting aperture through which the beam has passed. Typically, this value can vary from 0.2-1 for the HeNe laser, to as much as 10 for the higher powered pulsed laser systems such as ruby or Nd:glass lasers. For beams of uniform spatial distribution, the fraction $2/\pi$ in (3.8) is replaced by the constant 1.22.

As briefly mentioned in the introduction, a single-mode beam of laser radiation can be focused down to a spot approaching the wavelength of the light itself. In fact, the wavelength defines the fundamental limit imposed by diffraction theory, and one can actually achieve spots of this size by using the highest quality of focusing optics, and a good quality Gaussian laser beam. Hence shorter wavelengths can ultimately be focused to smaller spot sizes than longer wavelengths, and this in part explains the excitement of the recent progress in the performance of excimer lasers, and the thrust towards, for example, X-ray lithography. The actual radius w_0 of the focused spot can be calculated using

$$w_0 = f\phi \simeq \bar{F} \tag{3.9}$$

where f is the focal length of the lens, and F is the "F" number of the lens (F=f/d). It must be noted that both of these equations are relevant for the propagation of single-mode beams. For higher order mode structures the minimum spot size will be appreciably larger.

Again, the rules of diffraction dictate that a focused laser spot cannot remain so as it propagates in free space. Rather, it will diverge again as the beam propagates, as rapidly as it converged. The actual distance around focus where the spot size decreases from $w_0\sqrt{2}$ to w_0 and then increases to $w_0\sqrt{2}$, is known as the depth of focus, d_f. In laser applications where strong focusing is applied, and this could be either for small scale microstructures or for extended film formation, where a line focus is scanned over the substrate area, we must be aware of d_f as it indicates how much tolerance we can handle in the distance between the sample and the focus. The depth of focus is given by

$$d_f = \pm (\pi w_0^2/\lambda)\sqrt{\delta^2 - 1} \tag{3.10a}$$

where δ defines the increased spot size (in terms of w_0) that can be tolerated. For instance, if a 5% variation (i.e., $\delta = 1.05$) is acceptable then

$$d_f \simeq \pm w_0^2/\lambda . \tag{3.10b}$$

If an increase in spot size up to $w_0\sqrt{2}$ is tolerable ($\delta = 1.414$) then

$$d_f = \pi w_0{}^2/\lambda \qquad\qquad (3.10c)$$

defines what is commonly known as the Rayleigh range, or sometimes the confocal parameter, and this is usually defined as z_R. Clearly, d_f, is always a function of the square of the minimum spot size for a given lens. Therefore in a particular processing application, one must arrange a satisfactory balance between working distance, sensitivity of the process to fluctuations in this distance, and the dimensions required for processing on the workpiece.

For completeness, several formulae are included that are particularly useful for determining the properties of the Gaussian laser beam as it propagates. These relationships are very well known in laser circles and can be derived from basic optical resonator theory or by using the wave equation in homogeneous media [3.1,2].

At any point in space the Gaussian laser beam can be accurately and completely defined by two quantities, the spatial radius, and the radius of curvature of the phasefronts. Once known, the radius of curvature and spot size can be calculated for any other point in space along the direction of travel. The phasefronts, which are uniphase for the Gaussian mode, are spherical near the axis of propagation, and are defined by a radius of curvature, R(z), given by

$$R(z) = z\left[1 + (z_R/z)^2\right] \qquad\qquad (3.11)$$

where z is the distance along the axis of propagation from the previous beam waist. The radius of the propagating beam, w(z), is defined by

$$w(z) = w_0\sqrt{1 + (z/z_R)^2} \qquad\qquad (3.12)$$

105

where w_0 is the radius of the beam waist at the previous focus, and z is the distance from that point.

When applying these equations to a beam emerging from the laser cavity, it is important to locate the position and size of the beam waist formed by the cavity itself. Sometimes reference to the specification of the laser will be sufficient, but care must be taken to ensure that these characteristics are still valid. In any case, it is always possible to measure the actual spot size at various points along its axis outside the cavity and then fit (3.12) to the data in order to extrapolate to the effective values for z and w_0.

For single element focusing of the beam, it is well known that a thin lens of focal length f transforms the spherical wavefronts of the near axis Gaussian beam according to

$$1/R_1 + 1/R_2 = 1/f . \tag{3.13}$$

This gives the new radius of curvature of the beam emerging from the lens. It is usually assumed that the spot size is identical at the points immediately on either side of the lens. Since the radius of curvature of the beam has been altered by the lens, however, a new beam waist (real or apparent) will result. The precise position beyond the lens (z_2) where this will occur is given by

$$1/z_2 = 1/f - 1/\left[z_1 + z_R^2/(z_1 - f)\right] \tag{3.14}$$

where z_1 is the distance of the lens from the initial beam waist. (If z_R is infinite, such that the beam is parallel and non-divergent, the new beam waist would occur exactly at a distance z_2 = f from the lens). Now with a knowledge of the new value of R at a distance z_2 from the new focus, the new Rayleigh range (confocal beam parameter) can be calculated, whence w_0, $R(z_2)$ and $w(z_2)$ can also be found, using (3.10c,11,12). A more elegant routine for transforming a Gaussian laser beam through lens-like media involves the use of the so-called ABCD law [3.1a].

3.2 Spatial Resolution

We have seen already that laser radiation can, with the use of the highest quality "diffraction-limited" optical components, be focused down to a spot size approaching its wavelength. Thus with the use of UV laser radiation it is possible to achieve illumination diameters over some depth of focus better than 0.3 μm. Indeed, the use of monochromatic laser radiation as a substitute for conventional incoherent sources of radiation in the usual lithographic processes of the semiconductor production line is currently being considered in the quest for improved optical resolution. Although the wavelength is an important consideration here, the actual intensity of the light source is also advantageous. Incoherent systems are currently being pushed so far into the UV that exposure times of the order of 30 s are not uncommon for photoresist materials. Such long illumination times force one to consider stability against external vibrations during the process, but intense laser radiation enables similar exposures to be taken in fractions of a second, thereby eliminating such problems.

However, it is not solely the spot size that governs the final linewidth achievable in the process. One must remember that the illumination of the material is only the first step in a series of processes that give rise to a delineated pattern. Conventional photolithography relies on the photo-induced chemical transformation of a thin layer of photoresist material. Resists typically contain three essential components: a polymer base resin, a photosensitive ingredient, and a volatile solvent. During irradiation, the unsaturated bonds in the negative-type resists form longer or cross-linked molecules, while in the positive-type resists the saturated bonds are broken. Following exposure to the radiation, the material is then developed in such a way that the required areas of the film that were irradiated (positive) or unexposed (negative) are selectively removed in an etching reaction. The photo-sensitizer and the incident wavelength of the radiation must clearly be chosen to maximize the efficiency and resolution of the process. In conventional optical lithography, however, wavelengths shorter than about 300 nm are not used because there are presently no known effective combinations of resist and optical wavelength.

In laser direct writing, there will also be contributory factors other than the spot size that will determine the ultimate resolution of the system, and these will depend on the particular mechanisms involved in the structural formation. For example, for processes that will be thermally driven, the laser induced temperature gradients will dictate the rate of the chemical reaction across the sample surface. This will depend upon the spot size, the intensity distribution within the spot, the optical and thermal properties of the material and of the ambient, and the characteristics of the irradiation event itself with respect to scan speed, irradiation time etc. Optical absorption processes that change during irradiation, as well as multiple interference effects in the thin films, will also inevitably cause some distortions to the ultimate resolution, as will the nature of the chemical mechanism behind the delineating process itself.

An example of this effect on resolution can be seen by considering laser-controlled gas phase deposition. In this case, the linewidth of the deposited layer will be determined by the mean free path of the laser-excited particles. After dissociation or excitation to a reactive state, the molecules can diffuse in the gas phase before reaching the required surface sites, and the amount of diffusion should be minimized in order to obtain optimum resolution. The mean free path, and ultimately the linewidth, could be reduced either by increasing the total number of particles in the gas phase, or by increasing their collisional cross-sections. This can be achieved by operating the reaction with higher concentrations of reactants, or indeed of the buffer gas, or simply by increasing the temperature or pressure of the environment. Clearly there will be an optimum set of conditions for every set of laser induced reactions, but generally all of these remedies should work to some extent within limits, before undesirable effects become apparent.

EHRLICH and TSAO [3.3] have discussed the effects of these various nonlinear mechanisms on the ultimate resolution of a photoresist-based process and a microchemical process, each initiated by a laser beam. Expressions for the minimum resolvable linewidth L_0 normalized to the width of the laser beam, $2w_0$, for the cases of patterning photoresist, (3.15), and a typical Arrhenius-controlled process, Eq.(3.16), have been given as

$$\frac{L_0}{2w_0} = \frac{\log(2) + 1/\gamma^*}{\sqrt{2\log(e)}} \qquad \text{and} \qquad (3.15)$$

$$\frac{L_0}{2w_0} = \sqrt{\frac{1}{2\log(e)\,\gamma^*}} \; , \qquad (3.16)$$

respectively, where γ^* is defined analogously to γ describing the contrast factor of a photoresist [3.3]. In the present situation $\gamma^*(I_p)$ for a reaction described by an Arrhenius behaviour, and a linear relationship of (laser intensity/temperature change) near the operating temperature, is given by

$$\gamma = \frac{\ln(10)TE_a}{k(T + T_s)^2} \qquad (3.17)$$

where E_a is the activation energy of the process and T_s is the ambient temperature.

A close study of Fig.3.2 indicates that while the minimum resolution within the photoresist saturates with increasing γ^*, e.g. a value of approximately 0.6 at $\gamma^* = 50$, the microchemical

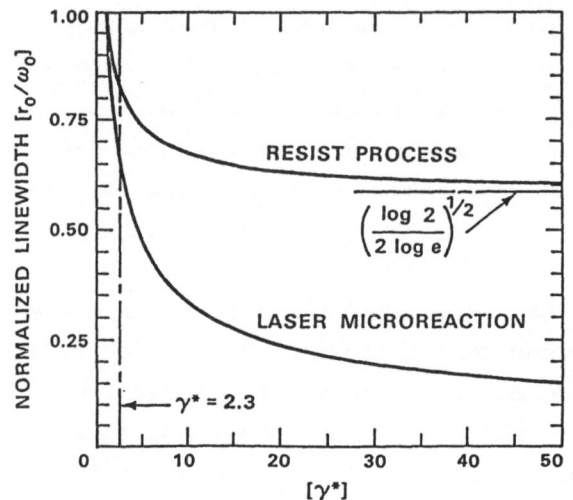

Fig.3.2. The minimum resolution available with photoresist in terms of normalized line width against γ^*, is compared with line widths available using laser induced microreactions [3.3]

linewidth falls steadily to around 0.15 at the same value of γ^*. Clearly there will be an optimum range of γ^* where the resolution will be carefully balanced against the stability and repeatability of the reaction, but at typical values of γ^* between 20 and 50, linewidths up to 1/4 of those achieved with a photoresist can be obtained. In fact, factors of around 2.5 reduction from the wavelength of the radiation used have already been achieved by PODLESNIK et al [3.4] during etching of GaAs, and by EHRLICH and TSAO [3.5] during deposition of poly-crystalline silicon, where the incident spot size was nearly one order of magnitude greater than the linewidth obtained.

A final example of this effect is shown in Fig.3.3. As the theory in Sect.2.5 indicates, the temperature profiles induced by some laser beams can remain quite close to the original dimensions of the heating laser beam. If a reaction such as silicon oxidation is induced on the surface and is considered to be totally controlled by the temperature of the substrate, then the temperature profile shown in the figure will give rise to the sharper feature of oxide growth. By carefully calculating the growth rates and times, and with the aid of conventional acid etching to reduce some of the less desirable background oxide, even more extreme profiles could be obtained.

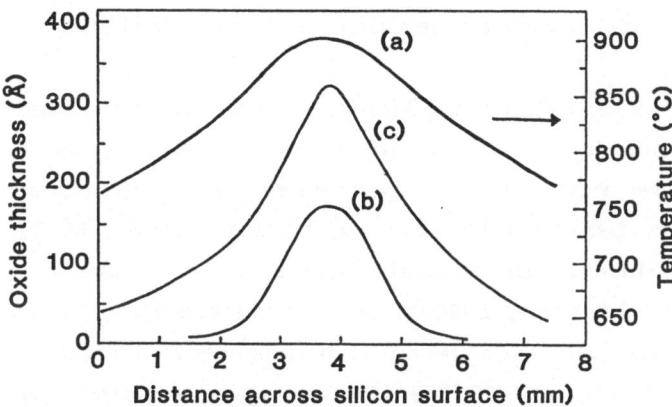

Fig.3.3. Illustrative example of the relative extent of the temperature profile (a) induced by a heating Gaussian laser beam (b), and the local growth of a thin film, in this case silicon dioxide

Although the laser radiation can be optically confined to dimensions less than 0.3 μm, many other processes could dictate the ultimate size of the processed area. A highly nonlinear reaction can assist in improving the ultimate resolution. However, material diffusion, as well as heat diffusion can be detrimental to this cause. In the case of heat diffusion being a problem one could consider operating under conditions where higher peak temperatures could be repeatedly induced for times shorter than the thermal diffusion times, with a significant "cooling-off" period between processing cycles. There may of course be many instances in thin film preparation where the resolution limits discussed here are not necessary, e.g. in large area depositions, where properties such as beam uniformity and energy stability are appropriately more important. These characteristics will be discussed in the next section.

3.3 Modes of Laser Processing

3.3.1 Geometrical Configurations

Conceptionally, there are many different modes of operation for laser processing. However, two separate classes of process can be isolated, depending upon whether one requires localized or large area coverage.

Several aspects of localized applications have already been addressed in the previous section in terms of the ultimate resolution available with various processes. In this mode, microphotochemistry, pantography [3.6] or direct writing [3.7] can be performed. This concept of maskless patterning, however, did not originate in the early 1980's, nor during the mid-1970's when laser annealing was discovered, but has been discussed over the years since the mid-1960's. For example, a patent of SOLOMON and MUELLER [3.8] in 1968, described the use of a directed laser beam focused on to a sample of ceramic, metal, or silicon, to raise the temperature sufficiently to initiate certain pyrolytic chemical reactions in the presence of particular vapours. They also discuss possibilities for complex relief definition by means of movable stages. Since then, of course, these concepts have been applied in many areas such as laser cutting, welding, scribing, and trimming of a variety of materials.

The user usually has two choices for pattern definition. The sample can be moved by means of a programmable motorized stage, enabling positioning of the workpiece with respect to the laser beam in two, and sometimes three dimensions. Here, one has limitations in terms of speed of processing, accuracy of alignment, and resolution. There is presently burgeoning activity in the development of positioning equipment capable of resolution in the region of 0.1 μm.

Another possibility for direct writing applications is to keep the sample stationary and to move the laser beam across the surface of the workpiece. Since lower loading of the motors is involved here, and angular movements can be employed, processing speeds can be significantly faster. There are also additional advantages in that the sample can be held in position very easily by vacuum stage, and can even be conveniently preheated without disturbing the positioning capability of the system. On the other hand, the optics required in this case invariably need to be quite complicated, particularly in applications where spot size stability is required. The problem involves the angular nature of the scanned incident laser beam, which will inflict small changes in the focused spot size on the sample, unless specially corrected flat-field lenses are used. Clearly, this problem becomes even more severe as one attempts to cover increasingly larger areas on the sample. Although the development of special optics to handle this situation is also currently being pursued, there is a general consensus of opinion that sample movement is preferable. The majority of investigations to date have preferred the optically simpler approach of using specially coated (for the ultraviolet) microscope objective lenses to achieve the smallest spots possible, and linear x,y positioning equipment. One of the more comprehensive experimental set-ups recently employed by McWILLIAMS et al. [3.6], is shown in Fig.3.4.

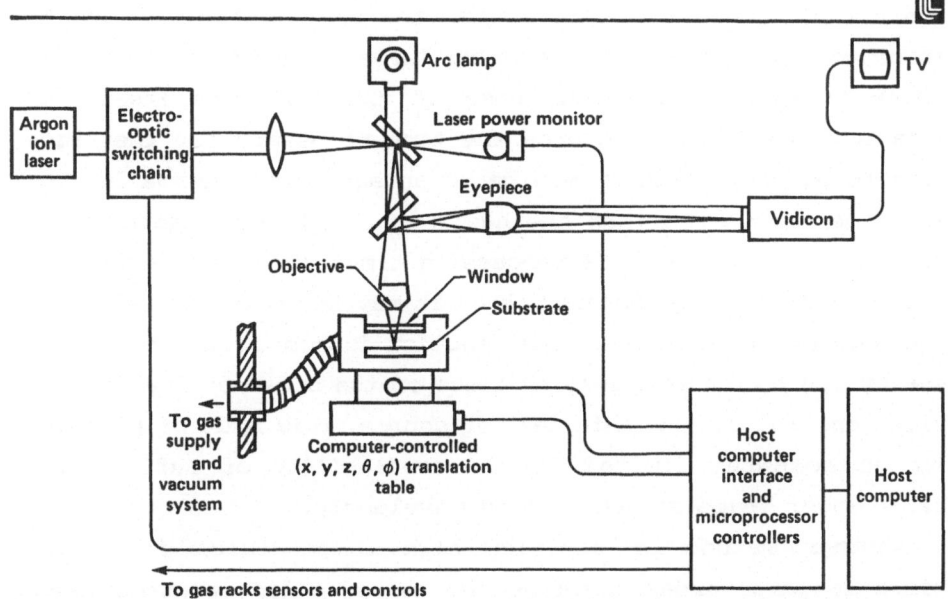

Fig.3.4. Schematic of origional experimental arrangement initially used by McWILLIAMS et al. [3.6] at Lawrence Livermore National Laboratory, for direct writing by laser

The essential elements of this application include an argon laser with the capability of temporal modulation, an optical microscope arrangement for simultaneous processing and monitoring, real-time beam characterization apparatus, and the processing chamber mounted on a computer controlled (x,y,z,θ,ϕ) translation stage. Clearly, the laser wavelength must be chosen to suit the requirements of the chemical reaction, and the accompanying option must enable sufficiently lossless transmission of the beam through the system to the sample chamber. The basic framework of this set up can obviously be duplicated to perform similar operations using different gas mixtures and different lasers. Experimental configurations such as this one, are usually required for localized operations such as etching, film deposition or growth, and in general any discrete evaporation, repair, or alloying application.

For larger area coverage, several alternative geometries are available, but these are much more dependent upon the type of reaction necessary. For example, for pyrolytic reactions, the laser radiation must impinge upon the sample surface, and be absorbed. One approach to covering large areas is to use a beam whose dimensions approach those of the workpiece itself. In this instance, depending upon the time necessary for the reaction to be completed, it may be advantageous to thermally isolate the sample, so that the heat generated in the material is used most effectively. If processing times are short, such that pulsed radiation can be used, then it may be useful to heat sink the sample, so that once the reaction is completed the excess energy can be drained away in times of the order of the thermal time constant of the material. Of course, this mode of processing is somewhat limited to the availability of sufficiently large laser beams at the desired wavelength.

Another method of applying high temperatures for short times to large areas involves the use of a line-scanned laser beam. In this way, a circularly symmetric Gaussian laser beam is transformed into an eccentric elliptical beam which is swept

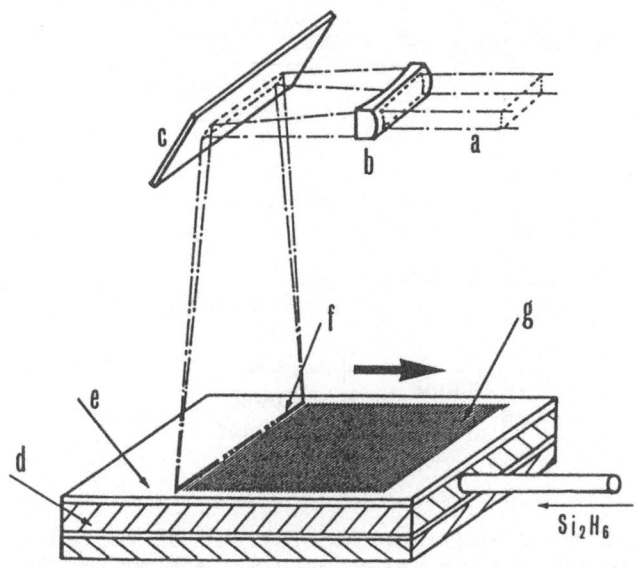

Fig.3.5. A line-scan system for laser processing, employed by MURAHARA and TOYODA [3.9]. (a: excimer laser, b: double cylindrical lens, c: mirror, d: movable gas cell, e: quartz window, f: linear focussed laser beam, g: deposited film)

across the workpiece in the direction of the minor axis of the beam. The dwell-time of the laser beam can be controlled by the width of the "line" beam, and the speed with which it is swept across. Again, a heat sink can be introduced. Figure 3.5 shows an application of a line-scanned laser beam for large area deposition [3.9]. On the other hand, if high temperatures are needed for extended periods of time, the line scan rate can be speeded up, and the heat sink removed in order to thermally isolate the sample and reduce heat losses. This is the regime of isothermal processing. Rapidly scanned spots can also be used in this way, but because of restrictions on the scan speed, and the natural thermal time constants of solids, it is not usually possible to cover large enough areas.

In the regime of photolytic processing, the laser beam need not be incident upon the surface of the material, especially when deposition from the vapour phase is required. All that is usually necessary is that the photoexcited species are gener-

115

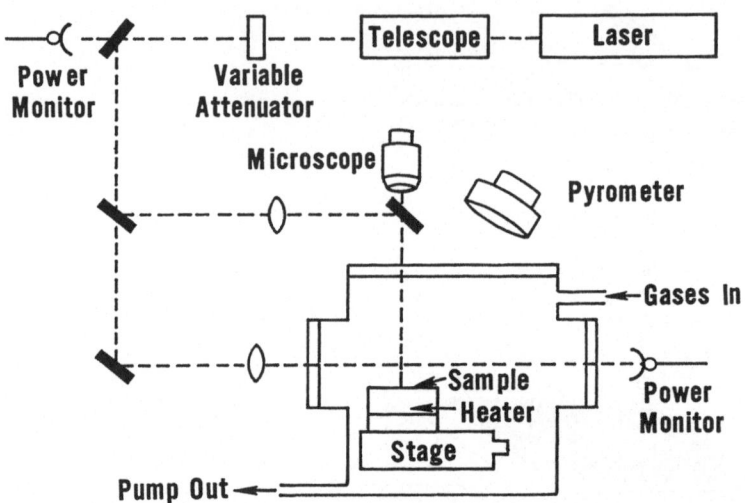

<u>Fig.3.6.</u> An example of a common experimental set-up for sur-
face processing of thin films with laser beam(s). Either of the
two beams can be used alone, or they can be used together in a
novel fashion to stimulate original processing conditions

ated close to the surface. This of course does not preclude the
use of any of the geometries described thus far, which for many
photolytic reactions may be quite desirable. Figure 3.6, however, shows schematically, the type of experimental configuration that has been used most often for the deposition of thin
dielectric films on silicon [3.10]. The laser beam that controls the reaction is directed across the substrate surface,
but some millimetres above it. The proximity of the beam to the
sample ensures that the products of the photostimulated reaction condense and nucleate on the material in that neighbourhood, and do not inefficiently coat other parts of the chamber,
as is the case in the conventional photodeposition. It is often
necessary, however, to pass a continuous flow of inert gas
across the window of the chamber in order to eliminate
unwanted depositions in that region. With this configuration,
material is deposited along the length of the beam, and for
some radial distance away from its direction of propagation.
Sometimes a second beam is introduced into the chamber, incident from directly above the sample. It is known, for example,

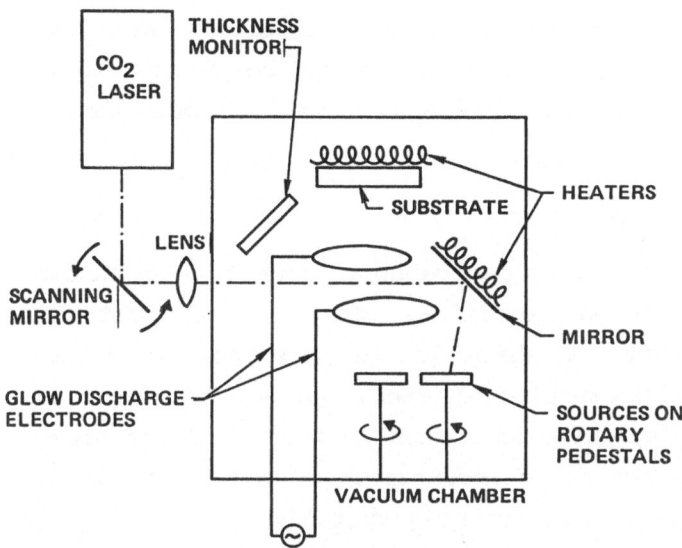

<u>Fig.3.7</u>. Sketch of apparatus applied to laser-assisted evapo-
ration of thin films, used by SANKUR and CHEUNG [3.11] at
Rockwell International Science Center

that the presence of this radiation within the substrate surface during many of the photodeposition reactions has a controlling influence on the physical properties of the deposited
film, as will be discussed later in Chap.6.

The arrangement used for laser evaporation (Chap.6) is
sketched in Fig.3.7 [3.11]. It basically consists of a modified
evaporation chamber, with the laser beam steering optics external to the chamber. The window is geometrically shielded from
the evaporated material, while the stationary mirror within the
reaction zone can be heated to limit the growth of deposits
during the reaction. It has been found that in most cases, any
thin coating that gradually forms is morphologically smooth
and does not absorb or scatter appreciable amounts of radiation. The evaporation sources used in these experiments were
typically pellets pressed out of 99.999 purity powder, and
these were placed over rotary pedestals. The substrates were
placed on heated stages capable of raising the temperature to
approximately 450°C, and the ambient pressure was maintained

at a level of 10^{-6} – 10^{-7} Torr during deposition. With only minor modifications the geometry can easily be changed to incorporate several sources, either in sequential fashion, or for simultaneous deposition.

3.3.2 Pattern Generation

The direct writing, or laser pantography concept is an innovation in microfabrication technology, and will no doubt find application in situations where limited drawing, correction, repair, linking or restructuring are required, or where low volumes (i.e. new designs, customized devices etc.) are necessary. It is presently quite inconceivable that a completely patterned dielectric, metal, or polycrystalline layer could be economically produced in this way for present day large-volume VLSI circuits. Furthermore, future devices will inevitably be even more densely packed. It is therefore becoming widely recognized that in these instances, large area pattern replication rather than line-by-line sequential definition will be essential. Indeed, the application of short-wavelength lasers has already been proposed as a replacement for conventional lamp sources in resist-based optical lithography [3.12], and is under serious consideration by at least one major company following recent research laboratory successes. Here, the few reported instances of laser-based pattern replication are summarized.

a) Background

There are three primary optical exposure techniques for pattern replication, known as projection, contact, and proximity. These are shown schematically in Fig.3.8. The simplest of these is probably the contact method, where, during exposure the mask is in physical contact with the sample, although they are temporarily separated during alignment. The intimate contact enables very high resolution to be achieved, but this method suffers from damage due to impurities becoming lodged between the mask

118

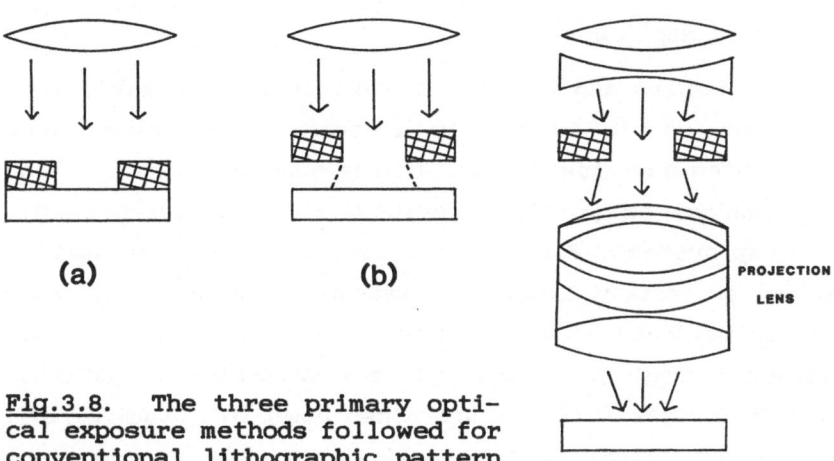

(a)　　　　　(b)

PROJECTION
LENS

(c)

Fig.3.8. The three primary optical exposure methods followed for
conventional lithographic pattern
generation

and sample. Proximity exposure minimizes but does not entirely
eliminate damage, by introducing a small (around 25 μm) gap
between mask and workpiece. Whereas 1 μm resolution is possible
in 0.5 μm of positive resist with contact exposure, the resolution for proximity printing is around 2-4 μm, given by $\sqrt{\lambda d}$,
where λ is the exposure wavelength and d is the gap distance.

In projection printing, an image of the mask is directed onto
the sample by means of a high resolution lens system placed
between the mask and the sample. In this way, contact or proximity damage is virtually eliminated, the mask lifetime being
limited now only by handling techniques. A one-to-one projection system, where a complete wafer is covered with the mask
image in a single exposure, is capable of resolution of around
2-3 μm and an overall registration accuracy that can be less
than 0.5 μm. In this system the mask and wafer are of similar
dimensions. It is now increasingly more common to find so-
called step-and-repeat, or direct-step-on systems, where the
mask may be up to 10 or 20 times larger than the projected
image, and only a portion of the sample is exposed at one time.
By stepping the wafer, and sometimes 81 steps are used, complete coverage is obtained, with resolution in the 1 - 2 μm
range, and registration accuracy better than 0.5 μm.

119

b) Laser Assisted Replication

The high peak powers and very short wavelengths available from excimer lasers enable previously unattainable microreactions to be stimulated and this has led to the consideration of these radiation sources as possible replacements for the commonly used arc-lamp systems. For high resolution damage-free pattern replication, projection imaging is probably the most realistic approach, as outlined above. However, early work in this field used contact lithography apparatus to etch submicron patterns in various materials with feature sizes down to 0.3 μm [3.13]. HENDERSON et al [3.14] used stencil masks made by electron-beam lithography in a similar contact mode, but in conjunction with an F_2 excimer laser (157 nm) reproduced etched features as small as 200 nm in self developing resist. CRAIGHEAD et al. [3.15] and WHITE et al. [3.16] similarly used the F_2 laser to reproduce mask patterns as narrow as 150 nm in conventional resists. These are so far the smallest features yet delineated by "conventional" photolithography techniques.

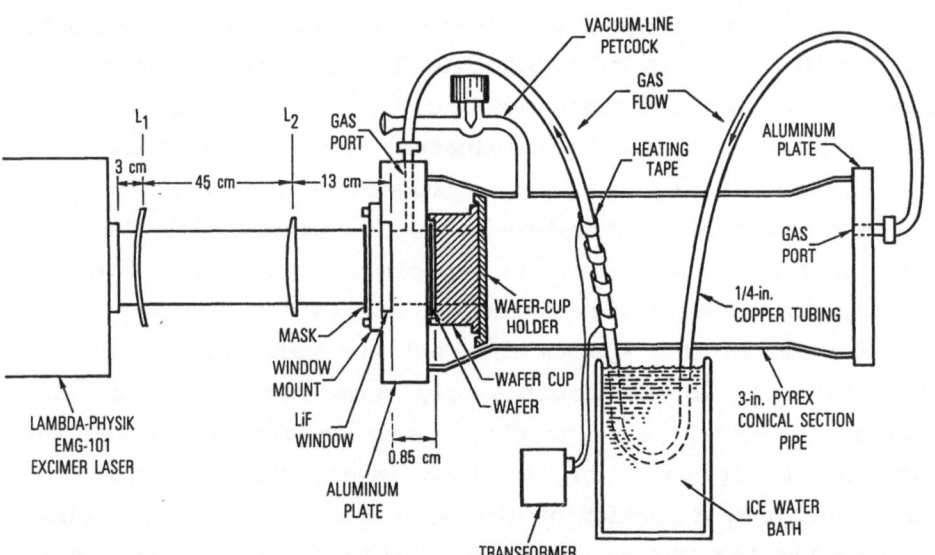

Fig.3.9. Scheme of the experimental arrangement used by LOPER and TABAT [3.17] at Aerospace Corporation for laser-induced radical etching studies

120

Another arrangement, more akin to proximity printing, was recently reported by LOPER and TABAT [3.17], and is shown in Fig. 3.9. Here, the wafer is mounted some 85 mm behind the mask, and it is illuminated at normal incidence by the laser beam. Figure 3.10 shows an early Nd:glass laser projection system used by METEV et al. [3.18] in 1980 to selectively oxidize chromium. Incidentally, it is interesting that this technique was used to fabricate masks which were subsequently used in future projection systems. Chapter 4 discusses more details of this work.

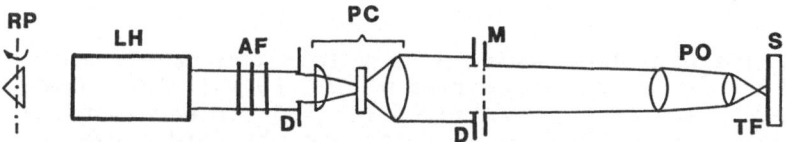

Fig.3.10. One of the earliest reported laser projection systems for pattern generation, comprising the following elements (RP: rotating prism, LH: laser head, AF: absorbing filters, D: diaphragm, PC: phase corrector, M: Mask, PO: projection objective, S: sample, TF: thin film) used by METEV and coworkers [3.18] at Sofia University

Projected excimer laser beam induced microchemistry has also been reported by LATTA et al. [3.19] and EHRLICH et al. [3.20]. The four-element projection lens used by LATTA et al. is shown in Fig.3.11. With an ArF laser, they achieved feature sizes as small as 2 μm in various photoresist materials of the order of 1 μm thick, although the normal design resolution of the lens was 10 μm. EHRLICH et al [3.20] describe projection imaging with both the ArF and F_2 lasers, achieving well-resolved 0.4 μm lines and spaces. A useful resume of the present state-of-the-art is also presented in the same reference, with details on the use of a Schwartzchild reflector objective, and various condenser optics. The projection techniques were applied to exposure of multilayer organic resists, doping of surfaces in reactive vapours, etching of glass, and solid phase transformation of Al/O cermets. It is clear that this field now

121

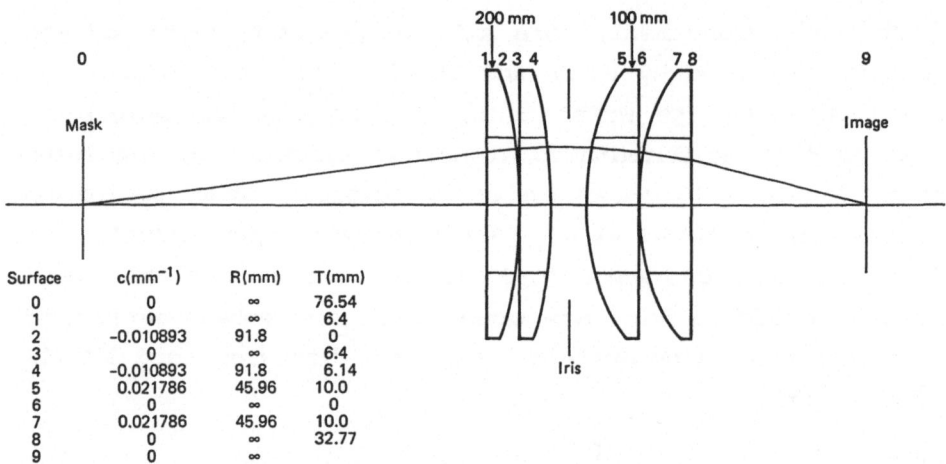

Surface	c(mm⁻¹)	R(mm)	T(mm)
0	0	∞	76.54
1	0	∞	6.4
2	-0.010893	91.8	0
3	0	∞	6.4
4	-0.010893	91.8	6.14
5	0.021786	45.96	10.0
6	0	∞	0
7	0.021786	45.96	10.0
8	0	∞	32.77
9	0	∞	

Fig.3.11. Projection lens made up from four quartz elements, as designed by LATTA and colleagues at IBM [3.19]. The curvature C, radius R and thickness T are indicated for each surface in the accompanying table. The elements are standard lenses of nominal focal length 200 and 100 mm measured at a wavelength of 589 nm

requires the development of reactive material systems capable of taking advantage of these short pulses and short wavelengths.

3.3.3 Process Uniformity and Reproducibility

Whether localized or large-area processing, it is crucial that the working parameters are stable during the reaction, and also reproducible on the next occasion. In terms of material properties, this requires that the optical and thermal properties are uniform over the entire area to be processed, and also from sample to sample. In materials where these characteristics are highly nonlinear, of course, this becomes almost an impossible task suggesting that unless some form of positive feedback is applied during the processing event the operation could not be easily transferred to production line status.

With respect to the properties of the laser radiation, it would be desirable to obtain a stable and repeatable beam power, spot size and mode profile, temporal duration, and wave-length. Presently, the shot to shot reproducibility of the best Q-switched lasers, is better than 5%, but closer to 20% for most high power mode-locked lasers. On the whole, the cw lasers can produce a much more stable output, certainly to within better than 2%, especially when various feedback mechanisms can be applied. Although it is not clear at the moment as to the optimum uses of the high intensity mode-locked laser systems in materials processing, it is possible to regulate the energy output from these lasers by means of a recently developed photonic regulator [3.21]. The device offers smoothing of the laser output on a picosecond timescale, is completely passive, and utilizes, coincidentally, the nonlinear optical absorption and self-refraction in silicon to operate.

The nature of the beam profile can often be a problem in many laser systems. Unless there is some form of damage to the optics, or key items within the cavity have been replaced or moved, it should always be possible to regain a Gaussian beam profile in a commercial laser system that has specified such an output. It is often advisable to obtain a clean output at the source of the laser. Nevertheless, as often happens once the beam has encountered a few optical components of less than optimum quality, and certain types of filters and prisms often fall into this category, the beam takes on a less than desir-able spatial profile. Should one require a clean Gaussian pro-file for processing at this stage, then it becomes necessary to spatially filter the high frequency noise from the beam. It has already been mentioned earlier in this chapter that the TEM_{00} beam retains its form, even after transformations through vari-ous optical systems. By using a tightly focusing lens to focus the noisy beam, this will simultaneously Fourier transform the higher frequencies to points further out in the transform plane away from the focused beam. The insertion of an aperture of the appropriate dimension at this point, will only allow the cen-tral Gaussian beam to pass down the optical axis, while the noise will be effectively blocked. It is often necessary to include a second lens to rescue the cleaned-up beam from diverging rapidly out of control, but this is usually a straightforward process.

Although TEM$_{00}$ beams enable extremely fine resolution to be obtained, there may be applications where a uniform profile is more desirable, for example, in large area film growth. Alternatively, some laser systems such as the ruby laser are notorious for producing beams with very large intensity inhomogeneities across them e.g. hot spots, that are often a consequence of small differences in refractive index across the lasing medium, in this case the solid-state rod. Both of these situations introduce the need for a device to spatially homogenize the laser beams before processing.

CULLIS et al [3.22] have shown, by diverting a ruby laser beam into a quartz rod with a right angle bend and a rough entrance surface, that sufficient multiple internal reflections are generated to mix up the spatial and temporal intensity and phase of the original beam, and present at the polished exit surface, an essentially uniform (spatially and temporally) beam profile (Fig.3.12). Further modifications to this original

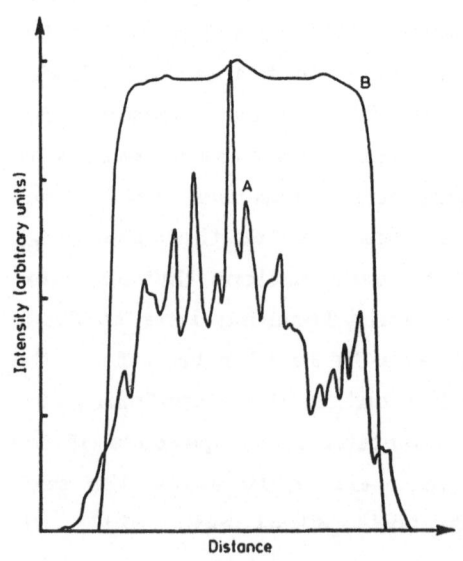

Fig.3.12. Homogenized intensity profile (B) of an initial multimode pattern (A) obtained using a quartz diffuser patented by CULLIS et al. [3.22] from RSRE in 1978

124

design have appeared over the years, and the better efforts have given a beam uniformity of approximately 5% in fluence over an area almost equal to the output face of the quartz waveguide. Perhaps the biggest drawback of this method is the rapid divergence of the outcoming beam, which always demands that the sample be placed within a few millimetres of the output face of the homogenizer. The spatial uniformity, of course, is also a strong function of position behind this exit face. GROJEAN et al. [3.23 have also designed a device based upon the principle of the kaleidoscope that is capable of homogenizing the profile of a CO_2 laser beam to within 8%, with an overall energy loss of only 14%. One can also envisage the use of a hollow polished metal tube, similar in shape to the quartz rod designed by Cullis et al., which could also be used for infrared laser beams.

Several alternative approaches to transforming the non-uniform intensity distribution into a more uniform profile have been proposed since the initial publications described above. LACOMBAT et al. [3.24], and later KAWAMURA et al. [3.25] reported on the use of wedge prisms, while a patent by BRUNSTING [3.26] and an optical integrating device manufactured by SPAWR Optical research [3.27] use segmented mirrors to achieve the necessary uniformity. However, both the prism-, and the mirror-based devices, which can comprise up to 25 segments [3.27], provide a uniform beam by optimized overlapping of the individual segments only at one single focal plane in the system.

LATTA and JAIN [3.28] have developed an improved version of the folding mirror system by using a wedged tunnel configuration. With this technique, not only was a superior uniformity produced, but collimation was also achieved. With a two-stage device, an intensity uniformity of approximately ± 1.5% was achieved over an area containing around 75% of the beam energy.

One possible drawback of these devices is their effect on coherent beams. They either scramble the coherence totally or cause interference fringes to develop when monomode beams are

125

used. However, we have already seen that coherence is not a necessary prerequisite for laser chemical processing, and so this loss of coherence will rarely be detrimental to the process. In fact, it may even be beneficial under some circumstances. There have occasionally been unique problems associated with the spatial uniformity of films deposited or grown by laser initiated mechanisms, in terms of film thickness or stoichiometry. This is known to be due to "radiation remnant waves" propagating across the surface of the irradiated material during film growth, leading quite often to periodic energy deposition and the formation of so-called "ripples". These have been observed in many materials (semiconductors, metals and insulators) under laser illumination [3.29-34], and have led to the development of a "universal" model based upon electromagnetic scattering by the natural roughness of the sample surface [3.35,36], although there still remain reports of apparently fundamentally different "ripple" phenomena [3.37]. The orientation of these ripples is often intimately related to the direction of the dominant electric field component of the incident radiation, and therefore one possible approach to minimizing their formation would be to apply optical energy with no preferred electric field direction, such as circularly polarized radiation, or light with a short coherence length. Most diffusers will produce beams with the latter qualities, thereby assisting in the battle against ripples.

Finally, we mention the patent of KREUZER [3.38], and publication of SHAFER [3.39], which employ specially designed lenses to redistribute the peak central intensity of the laser beam around the edges of the profile. Here, unlike in the applications mentioned above, the phase of the wavefronts is not disturbed, and spatially coherent monomode beams may be homogenized.

3.3.4 Beam Profile Measurements

Whilst the absolute power (energy) content of (a pulse of) laser radiation can be measured quite accurately with, for example, a thermopile or calorimeter, there are often problems pertaining to the absolute spatial dimensions of the laser beam. This leads to difficulties in defining fluence thresholds and related quantities, and ultimately to problems with the analysis of the kinetics of particular reactions. In recent years, this problem of measuring the spatial profile of a laser beam has been solved to a satisfactory degree, albeit at cosiderable expensive, by means of vidicon and reticon detectors, and also photodiode arrays which can be used to detect and store sections or even the complete two-dimensional extent of laser beams at various wavelengths.

A more traditional, and less expensive method involves the translation of a pinhole (down to around 5 µm diameter are commercially available) and detector of appropriate sensitivity, across the beam at the point where the sample would be situated. Higher power beams cannot be characterized this way because of the problem of damaging the pinhole by initiating undesirable chemical and physical processes (a case of unwanted laser induced chemistry). Over the years alternatives methods have included the use of intervening beam choppers with large mark to space ratios, or large numbers of optical filters in the beam path to reduce the absolute power incident on the detector. Burn marks on certain types of materials can also give an indication of spot sizes, while fluorescent screens can be used to obtain a reasonable estimate of IR beam dimensions.

For intense CO_2 laser beams, two approaches involving the use of a ball-bearing [3.40,41] and a rotating wire [3.42] have been reported. These reflect small fractions of the beam on to a conventional detector, and by moving one with respect to the other, or indeed the beam across the system, a complete profile can be recorded without destroying the detector. Once the profile has been obtained, and the radius accurately determined, the fluence of the radiation can be precisely calculated. Remembering that $F = E/\pi w^2$, any error in the knowledge of the beam radius will result in a doubled error in the calculated fluence.

3.4 The Choice of Laser

Lasing action has been observed in several hundred different mixtures of active media over the years, giving rise to a great diversity of available wavelengths and pulse durations for the laser user. This wide choice in the colour of the radiation, however, has mostly benefited spectroscopists. While these studies are essential to the fundamental understanding of many beam–solid interactions, only a few of the laser systems available have actually been adopted for materials processing. Table 3.1 summarizes the main group of lasers that have been regularly applied to the processing of materials in the past few years.

Generally the wavelength criterion is based on a weighted product of the amount of power available, and the optical and thermodynamic properties of the material to be processed, as well as the nature of the reaction mechanism required. Most dye lasers, or the new generation of ultrashort pulsed lasers, either do not produce sufficient energy or are such complicated systems that their integration into any present–day mass production line is too difficult to consider. Besides, it is not clear how 8 fs pulses, or some of the wavelengths available with some dye lasers will find useful applications in this field at present.

If the process under consideration is thermally controlled, then radiation must be strongly absorbed by the sample so that the induced temperature increase is large enough that the reaction can proceed at the necessary pace. In this case, the exact choice of wavelength would not be critical, because of the band structure of solids (Chap.2) and there would usually be a number of laser systems to choose from. However, one must also be aware of the optical reflectivity of the workpiece; the absorption at a particular wavelength can be very high, but so can the reflectivity, and most of the radiation incident on the sample could actually be reflected away. However, as shown in Fig. 2.5, the reflectivity of many metals varies by only a factor of 2 or 3 over the entire spectrum from 300 nm to beyond 20 μm. The absorption, however, can vary by several orders of magnitude over the same spectral range. The power available from commercially available laser systems also varies by orders of magnitude over this same range, the IR systems being traditionally the most powerful. Therefore it is logical that most metal processing is performed by high power CO_2 lasers, which operate at a wavelength where absorption is, without exception in metals, extremely efficient.

Semiconductor processing that is thermally dominated can, on the other hand, be most readily effected by radiation whose photonic energy is greater than the bandgap energy of the material. For example, laser heating of silicon (band gap $\simeq$ 1.12 eV) has been largely investigated using argon ion, ruby, frequency doubled Nd:Yag, and excimer lasers, all of which operate in the visible/UV region of the electromagnetic spectrum (E = $h\nu \geq$ 1.8 eV). This criterion, it must be noted, does not in any way exclude the use of other wavelengths and laser systems for this application. However, the photon energies available from the other systems would require nonlinear optical absorption processes to effect the required heating, and this would make the overall processing less easy to control, and consequently less desirable on a production line.

When the process depends on the breaking of chemical bonds, more energetic photons are usually required, and therefore, UV lasers are commonly employed. It is also possible to use IR lasers in this regime to encourage multi-photon absorption to initiate preferential photodecomposition. UV radiation can be generated by various nonlinear frequency conversion operations on longer wavelengths. In this way the one micron radiation

129

Table 3.1. Characteristics of the most commonly used laser systems in materials processing.

Laser type	Wavelength	Mode of operation	Average /pulse energy [J]	CW power [W]
Carbon dioxide (CO_2)	9-11	cw		15000
		pulsed (μs-ns)	100	
Carbon monoxide (CO)	5-7	cw		100
		pulsed (μs)	0.5	
Nd:YAG	1.064	cw		600
		pulsed (ns, ps)	50	
Nd:Glass	1.054	pulsed (ns, ps)	100	
Alexandrite	0.730-0.780	pulsed (ns)	5	
Ruby	0.694	pulsed (ns)	50	
Krypton	0.351-0.800	cw		8
Nd:YAG ($\lambda/2$)	0.533	pulsed (ns, ps)	10	
Argon	0.351-0.515	cw		20
Nitrogen	0.337	pulsed (ns)	0.01	
Nd:YAG ($\lambda/4$)	0.266	pulsed (ns, ps)	5	
EXCIMER XeF	0.351	pulsed (ns)	0.15	
XeCl	0.308	pulsed (ns)	0.25	
KrF	0.248	pulsed (ns)	0.75	
ArF	0.193	pulsed (ns)	0.5	

from the Nd:Yag laser can be converted to the UV by second harmonic generation to the green (533 nm) and further harmonic generation to the UV (266 nm) using special optically anisotropic crystals such as KDP. As Table 3.1 also shows, UV radiation is also available in low power cw form from Ar and Kr ion lasers, and in nanosecond pulses from nitrogen and excimer lasers. In photolysis the wavelength is a little more critical than in pyrolysis and a degree of frequency tuning can be applied to give a previously unattainable control on the overall chemical reaction. Usually, however, the absorption cross-section of the gas-phase molecules is appreciable over a significant range of wavelengths in the UV and the exact wavelength is not so critical as is commonly believed. (See for example Fig.2.3 for the absorption cross-section of several commonly used donor molecules and precursors for photochemical deposition).

The mode of operation of the laser is another important consideration when deciding which system to employ for processing.

For example, continuous wave lasers can be relatively easily controlled in terms of output power and stability, making it quite straightforward to regulate the incident flux and therefore the temperature increase in the sample in real time. On the other hand, it is much harder to consistently reproduce on a shot to shot basis, the same conditions of irradiation control for pulsed lasers. Furthermore, the optical interactions with most materials become much more nonlinear when shorter (and therefore more intense) pulse radiation is used.

There are certain advantages in some circumstances to using short pulses for materials processing. It is possible, for example, to deliver significant amounts of energy to the material before any can diffuse appreciably during the irradiation time of the pulse. An important consequence of this is that less energy is required to heat the sample to the desired temperature than would be the case with longer pulses. Also, the actual degree of heating that the remainder of the sample is subjected to would be quite minimal, and this has important implications in some areas of materials processing, as discussed frequently throughout this book. When heating is not the principal criterion, and one desires a photochemical reaction it is usually more advantageous to use a pulsed laser source. The extremely intense pulses can supply a sufficiently high density of incoming photons to initiate and sustain reactions even when the scattering cross-section for the absorption and photodissociation process is relatively low.

When the process and the substrate properties are known, it is most usual that only one or two types of lasers, if any, will suffice to initiate the reaction. Indeed the matching of laser wavelengths to currently available precursors is probably the strongest limiting factor in this field of study.

In very general terms, four main groups of lasers can be defined which have found specialist applications in the overall field of materials processing. The first, the CO_2 laser, is by far the most efficient high power laser available today, exhibiting conversion efficiencies from raw electrical power to coherent electromagnetic power at 10 μm of typically 20% and sometimes more. Consequently, it is possible to achieve system outputs in the multi-kilowatt range, and therefore in applications where large amounts of power are desirable, such as in cutting and drilling, this laser has found an enormous number of applications.

In recent years the excimer laser has been shown to produce radiation from the visible to the mid-UV with good efficiency, typically several percent [3.43]. Indeed the excimer laser is fast becoming the high power equivalent of the CO_2 laser in the ultraviolet region of the spectrum, and its development will continue in the coming years at a rapid rate. Through changes of the gaseous mixtures, wavelengths from around 351 nm to 319 nm, and lower, are presently available. In terms of photochemical reactions this laser seems destined to become the one most widely adopted. It must also be noted that the generation of much shorter wavelength coherent radiation is currently one of the most intensely investigated areas in quantum electronics, and the following few years may bring about systems capable of generation of wavelengths approaching 100 nm [3.44].

In applications where controlled melting and structural remixing are desired, the use of the frequency doubled Nd:Yag or the ruby laser has been popular. These lasers are usually operated in the Q-switched mode, giving rise to pulses of the order of 10 ns or so in duration. The Nd:Yag laser can be mode-locked, to give pulses as short as 20 ps, and can be used to pump a wide variety of dyes to give a larger variety of wavelengths. The ruby laser is the oldest laser system, having been first operated successfully in 1960, and emits radiation in the deep red region of the spectrum. Both of these lasers belong to the so-called solid-state lasers – the active medium being a rod of appropriately doped crystal, unlike the previously described systems which use gaseous mixtures as their active medium.

The final class of lasers described here are the noble gas ion lasers, such as the argon ion or the krypton ion systems. These operate in the visible, but can give rise to a wide variety of wavelengths from the IR down into the near UV, particularly the krypton laser. They usually operate CW, although they can be mode-locked if pulsed operation is desired. These systems have found applications in areas where controlled heating (without necessarily melting) is required, and have found particular applicability in semiconductor processing, where localized etching, oxidation, annealing, and even deposition have been studied. Also, because of the rather high energy content of the individual photons, there is a possibility that some nonthermal, or photo-initiated processes can also be induced by these lasers, as will be discussed later in Chap.4. These lasers are notoriously inefficient, typically of the order of 0.1%, or less, and require rather hefty power supplies, for example 40 A at 400 V, to supply some 10-20 W of optical power. Because of the nature of the device, in which the high current electronic discharge supplies energy to a confined gas, reliability has been suspect in the past, although modern materials development has improved this situation somewhat.

4. Laser-Assisted Oxidation and Nitridation

This chapter will review the laser-assisted growth of insulating films through controlled heating in the solid phase. Oxidation, being the most studied reaction in this area, forms the largest section of the chapter, although nitridation, and polymerization of thin organic films by laser induced thermal effects will also be discussed here. The tables in this chapter summarize all the oxide films formed by laser induced mechanisms that have been reported in the literature. The wider range of different techniques encompassed in these processes will be described in the following Chapters 5 and 6), which deal with insulator formation via laser induced phase changes, and laser assisted deposition. Our attention will not be restricted to dielectric layers for silicon technology, although this is where most attention has been focused; the formation of insulating layers on other semiconducting materials and metals, will also be discussed. These latter areas have enormous potential for advancing the expanding optoelectronic processing capabilities as well as localized surface hardening and protection for micromechanical applications.

4.1 Oxidation

The abrupt discontinuity of a material at the surface leads to the formation of many active sites that can form chemical bonds with various foreign atoms present in the vicinity. In this respect, many solid surfaces have a high affinity for oxygen atoms and consequently tend to eagerly and rapidly form thin oxide layers when exposed to air. In silicon, the bandgap electronic states present at a clean and abrupt surface, can be substantially reduced, sometimes by more than 5 orders of magnitude, by the presence of the thin surface native oxide. This remarkable passivating behaviour is one of the main reasons for the unparalleled success of silicon microelectronic technology. In some materials, this natural oxide layer can act as a barrier towards further oxidation by blocking the diffusion of the necessary species to the reacting interface. Where such diffusion is not hindered, oxidation can proceed according to the affinity of the material for oxygen, as well as the external environmental conditions for the reaction. In the following section, we will present details of some of the most successful descriptions of common oxidation reactions.

4.2 Background and Theory

Over the years a large number of oxidation laws have been formulated for many materials. For example, it is well known that composition of the gas mixtures, their partial pressures, the substrate temperature, and the thickness of oxide film already present on the surface of the material, greatly affect the reaction rate. The laws can represent logarithmic, linear, parabolic, linear-parabolic, cubic, or exponential behaviour, but in many cases, these are simple empirical formulations, and only vaguely relate to the precise nature of the reaction.

One of the most widely known oxidation laws is that of low temperature oxidation of metals, formulated in 1948/1949 by CABRERA and MOTT [4.1], and subsequently developed by FEHLNER and MOTT [4.2] in 1970. In the original model, which is similar in form to that used by VERWEY [4.3] for anodic oxidation, metal cations undergo interstitial diffusion between kink sites in the oxide already formed at the oxide/gas interface where they combine with the oxygen to form further oxide film. If W is the potential barrier at the first interface and an electric field F is applied to the bulk material, then the rate at which the charged species are formed, and therefore, the rate of reaction can be written as [4.1]

$$dx/dt = C \exp[-(W - qdF)/kT] \ . \tag{4.1}$$

In this case C is a constant, q is the charge on the ion, and d is the hopping distance between neighbouring interstitial sites. For low temperature oxidation the film must be considered to be transparent to electrons, which move either by tunnelling or by impurity conduction through defect states, such that oxidation is controlled by anionic rather than by cationic species. F is then replaced by V/x, where V is the potential set up between surface states at the oxide/gas interface and the bulk solid, divided by the film thickness.

For large x, where the factor $\exp(dV/x)$ in (4.1) is replaced by

$$\sinh(qdV/x) \simeq qdV/x \ , \tag{4.2}$$

(4.1) becomes parabolic in nature. Similarly, the Wagner relationship for oxide growth in thick layers can be written in the form [4.4]

$$dx/dt = (D_0/x) \exp(-T_0/T) \tag{4.3}$$

where D_0 and T_0 are constants.

As mentioned above, there are a number of oxidation laws that are applicable in particular conditions. We will now concentrate on one of the most commonly encountered oxidation laws, that of parabolic oxidation. In the formulation of this law, charged or neutral atoms or molecules are dissolved as defects, for example at interstitial sites, at either the gas/oxide interface, or the oxide/material interface. If the concentration at the interface is c giving a concentration gradient c/x, where x is the film thickness, the growth rate can be written

$$dx/dt = DVc/x \qquad (4.4)$$

where D is the diffusion coefficient for the defect, and V is the volume of oxide film grown when the defect reaches the reacting interface.

Upon integration of (4.4), we get

$$x^2 = 2DVct . \qquad (4.5)$$

At sufficiently high temperatures, and for thick enough layers, this equation describes the oxidation behaviour of silicon. Although this relationship had been known for the oxidation of silicon in this regime, DEAL and GROVE [4.5] were the first to formulate a satisfactory model for the high temperature oxidation of silicon which was valid over almost all film thickness. Silicon oxidizes very readily over a wide range of temperatures, producing an amorphous insulating structure that is known to be silicon dioxide (SiO_2), and forms the basis for a wide range of silica glasses. In 1965, DEAL and GROVE (DG) produced the celebrated formula

$$x^2 + Ax = B(t + t_0) \qquad (4.6)$$

where A and B are constants, and the quantity t_0 represents a shift in the time coordinate to account for the presence of a rapidly grown initial oxide layer of thickness x_0. In their model, the oxidant, e.g. neutral O_2 for dry oxidation, is transported from the gas phase to the gas/oxide interface, diffuses across any existing oxide layer, and interacts with the silicon atoms at the oxide/material interface. Solving (4.6) for x as a function of time gives

$$2x/A = 1 + 4B\sqrt{(t+t_0)/A^2} - 1 \qquad (4.7)$$

which is the well-known linear-parabolic characteristic of silicon oxidation. Two limiting cases can be defined. The first is the parabolic law, when $t \gg t_0$, written as

$$x^2 = Bt .$$

(4.8)

Here, B is known as the parabolic rate constant. The other limiting case occurs for much shorter oxidation times, when $(t+t_0)$ $\ll A^2/4B$, and defines the linear oxidation regime, i.e.

$$x = \frac{(t + t_0)B}{A} .$$

(4.9)

In this case the linear rate constant is the ratio of B/A. By plotting the growth of silicon dioxide in a convenient manner, as shown in Fig. 4.1a, both of these regimes can be quite readily identified. The reaction rate constants are more accurately defined by

$$B = 2DC^*/N_1$$

(4.10)

and

$$\frac{B}{A} = \left[k_s \frac{h}{k_s + h} \right] \frac{C^*}{N_1}$$

(4.11)

where $h = h_g/HkT$, h_g is the gas phase mass transfer coefficient, H is the Henry's law constant, C^* is the equilibrium bulk concentration of the oxidizing species, D is its diffusion coefficient, and k_s is the net rate constant of the chemical reaction at the interface, which may of course involve more than one reaction. Also, N_1 is the number of oxidizing molecules incorporated into a unit volume of the oxide layer. Inspection of (4.10) reveals that the parabolic oxidation rate depends strongly upon the diffusion coefficient, and therefore one refers to the reaction in this thick film growth regime as diffusion limited. In other words, as expected, the reaction

138

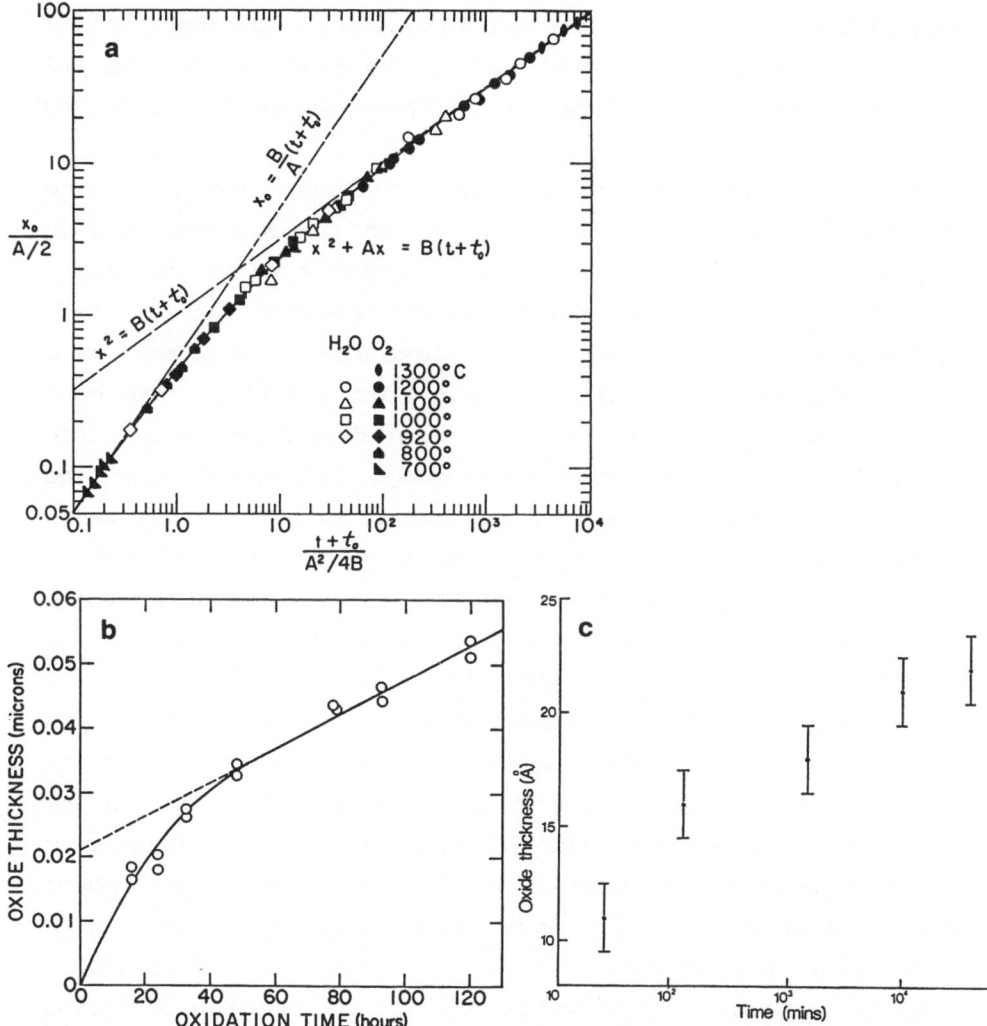

Fig.4.1. (a) Plot of oxide growth on silicon, indicating both linear and parabolic regime [4.5]). (b) Back extrapolation of the linear regime of the DG model for oxidation of Si at 700°C, indicating the necessity for an initial oxide thickness of some 20 nm [4.5]. The data show a gradual oxidation rate for this regime. (c) Oxide thickness on c-Si as a function of time after cleaning in 10:1 HF followed by a 15 min rinse in water and a 10 min spin dry in warm nitrogen [4.6]

only proceeds as fast as the oxidant can arrive at the reacting interface. Being proportional to C^*, the reaction rate should also be proportional to the partial pressure of the oxidizing species.

Examination of (4.11) reveals that the linear reaction has no obvious dependence on the diffusion of the species, and is strongly controlled by the actual reaction rate at the interface. In this case, the kinetics are described as rate-limited. Furthermore, the ratio B/A has the same linear dependence as B on C^*. This fact, that the linear and the parabolic rate constants should exhibit the same pressure dependence, was originally found for the temperature range 1000⁰–1200⁰C and the pressure range 0.1–1 atm [4.1]. However, in more recent years it has been shown that B/A follows more closely a power-law dependence on oxygen partial pressure, where $B/A = kp^n$, k being a constant and n lying mostly between 0.6 and 0.8 [4.7-9]. CAMELIN et al. [4.10] have recently found $B/A \propto P^{0.7}$, and $B \propto P$ between 600 and 780⁰C for pressures between 14 and 1000 atm.

In the parabolic regime, one might expect that the oxidation reaction should also reflect an Arrhenius dependence because of the strong influence of the diffusion coefficient. Indeed, it has been shown for the case of "dry" oxidation of silicon, where the trace amounts of H_2O were minimized to less than $1:10^6$, that the activation energy of the reaction was 1.23 eV [4.5]. This corresponds to the value of 1.17 eV for the diffusion coefficient of the oxygen molecule through fused silica. By the application of a corona discharge [4.11] at low temperatures to oxides previously grown at various temperatures up to 1000⁰C, the activation energy of oxygen diffusion through these films has been recently found to be only 0.8eV [4.12]. It is thought that the discharge can release any in-built stress in the oxide layers grown at temperatures below about 1150⁰–1200⁰C, where stress relief can otherwise be thermally obtained. The important role of stress in the reaction will be discussed in more detail later. For the oxidation of silicon in steam, i.e. "wet"

140

oxidation, an activation energy of 0.71 eV compares very favourably with the diffusivity of water in fused silica, which has
been experimentally determined to be 0.80 eV. Although almost
identical to the stress-relieved activation energy mentioned
above, there has not yet been any speculation of reaction similarities in the two systems.

In the linear regime where the chemical reaction (and not
diffusion) controls the rate, activation energies of 1.96 and
2.0 eV have been measured for the wet and dry oxidation reactions respectively, generally for temperatures above 700°C.
Because these two values are almost identical, it indicates
that the interface reaction may be the same in both cases.
These activation energies are in fact very close to the Si-Si
bond strength of 1.83 eV, and it is tempting to propose that
the reaction is limited by the availability of Si free radicals.
In this interpretation, however, there would be no allowance
for the energy required by the oxidant to dissolve into the
surface of the oxide. A slightly reduced activation energy of
1.7 eV has recently been found for temperatures between 600°
and 780°C [4.10] suggesting that stress generally introduced
into the oxide at these temperatures may weaken the Si-Si bonding close to the reaction interface. The contribution of stress
to the reaction has been suggested by several groups as will be
discussed later in this chapter.

Although the DG model has been universally adopted to
describe silicon oxidation, it is not without its anomalies.
While a fit of the theory to the data for wet oxidation of Si
extrapolates back to the origin, i.e. $x = 0$ at $t = 0$, thereby
defining $t_0 = 0$, an identical fitting procedure to the dry oxidation data leads to a value of x around 25 nm at $t = 0$. In
other words, the DG theory can only work for dry oxidation if
one assumes an apparent initial condition of $x \simeq 25$ nm. This
region of "anomalous" growth behaviour led some people to suspect that the growth of oxide in this regime might be instantaneous. Many independent studies [4.13-16] have now shown

141

that this is strictly not the case, although the reaction rate is indeed quite fast (see, for example, Fig.4.1b). Since in previous years, most oxides used in commercial applications were greater than 100 nm it became convenient to assume the presence of this "instantaneous" oxide thickness in order to accurately predict the oxidation rates. Consequently, the growth of oxides thinner than 25 nm did not receive significant attention. Present trends towards ever-decreasing device dimensions now require that high quality oxide layers of 20 nm or less, be produced, making it necessary to obtain a better understanding of the mechanisms underlying the formation of these ultrathin layers.

The oxidation rate of these very thin layers can be empirically fitted by a linear-parabolic curve, but with different rate constants than those extracted by DEAL and GROVE for oxidation of films thicker than 25 nm. Again another very thin oxide layer must be present on the surface to obtain a good fit. In this case, however, the initial oxide layer is real, being the well-known native oxide always formed rapidly after the freshly cleaned silicon surface is exposed to air. Typical native oxide thicknesses of around 1.5-2.5 nm are usually found on the Si surface (Fig.4.1c). The other constants extracted from the linear-parabolic trend of the data cannot be meaningfully applied to the original DG model in the sub 25 nm regime, and this has led to a proliferation of alternative oxidation models in the literature in recent years. In fact, together with the small but measurable deviations for the predicted pressure dependences as indicated above, this has brought into question some of the major assumptions within the DG model. Alternative approaches have considered the influence of more than one oxidizing species, or the possibility of several simultaneous reactions at the $Si-SiO_2$ interface and even the effects of strain and viscosity on the reaction. These, and many other published suggestions have been summarized in Table 4.1, and will be briefly reviewed below. The short-comings of the DG

142

Table 4.1. Summary of silicon oxidation models

DESCRIPTION OF SILICON OXIDATION MODEL	REF.
O_2 diffusion & reaction with Si+space charge effect & tunnelling/thermionic emission	4.5
Enhanced diffusion of O_2 through micropores	4.17,18
Stress-induced diffusion effects	4.19,20
Creation of O_2^-/hole pair, leading to enhanced diffusion of O_2^-	4.21,22
Influence of O_2^{2-} and O^-	4.23
Importance of O^-	4.24
Fixed charge in oxide assists diffusion charged species O_2^-, and perhaps O^-	4.25
Fixed charge in oxide reduces available Si bonds by reducing holes with band-bending	4.26
Both O_2 and 2-O react with silicon	4.27
Equilibrium between O_2 and 2-O, only O reacts with Si for thin oxides	4.28
Two-step oxidation: Si -> α-SiO$_2$ + Si(I) & Si(I) -> SiO$_2$	4.29

model were initially noted by the original authors, and they actually proposed that in addition to the movement of neutral oxygen molecules during dry oxidation, space-charge effects may be an important consideration during the early stages of the reaction, thereby enforcing ideas that charged species may be influential. In fact, as far back as 1962, JORGENSEN [4.30] found that an applied voltage could accelerate or retard the oxidation, although RALEIGH [4.31] subsequently pointed out that this result was ambiguous because of possible electrolytic effects due to the use of platinum electrodes. TILLER [4.32] has since shown that if electrons are available from an open

circuit, a mechanism involving the transport of anions is preferred. The work of MILLS and KROGER [4.33], and BARTON [4.34]
shows through electrochemical arguments that O^{2-} may be the
mobile species, because oxidation at 800°C can be stopped completely by applying a voltage of 1.78 V, while four times this
energy in electron volts per O_2 molecule is released during the
reaction. This suggests that four electrons must flow around
the circuit for each O_2 molecule to participate in the reaction.
Thus

$$O_2 + 4e \rightleftarrows 2O^{2-} \tag{4.12}$$

and the mobile species may be the O^{2-} ion. If ionic flow is
indeed appreciable, then the model of CABRERA and MOTT [4.1],
discussed earlier in this section, may be relevant here. This
model of course is dependent upon electronic charge flowing
through the oxide, by tunnelling, thermionic emission, or by
some other mechanism. It is possible that electrons could move
from the conduction band of the silicon to the conduction band
of the oxide (Fig.4.2 and Table 4.2), but only if they acquire an
extra energy of some 3.2 eV, and this is not considered to be
highly probable even at these elevated temperatures of oxida-

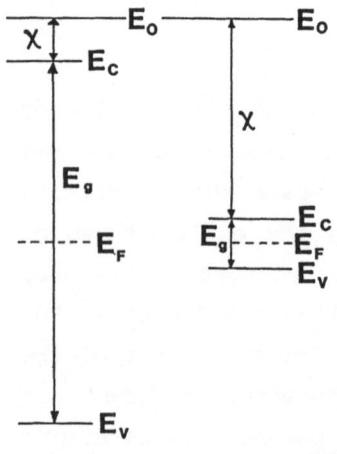

(SiO$_2$) (Si)

Fig.4.2. Schematic band structures at k=0
for c-Si and a-SiO$_2$

144

Table 4.2. Various constants related to the band structure of silicon and silicon dioxide

	ENERGY GAP E_g [eV]	AFFINITY χ [eV]	DIELECTRIC CONSTANT ϵ
Si	1.12	4.15	11.8
SiO_2	8	0.95	3.85

tion. Although such a mechanism has yet to be convincingly demonstrated in this regime, it may be important when the substrate is illuminated by sufficiently energetic photons, as in the case of laser induced oxidation, which will be discussed later in this chapter. The contribution of charged species in the reaction could be affected in a number of ways, for example, through diffusion, or the oxidation reaction itself. Unlike most metals, silicon oxidizes at the interface with its oxide and not at the oxide-air boundary. Evidence of oxidant diffusion across the SiO_2 to its junction with the silicon, rather than cationic transportation to the oxide surface, has been obtained over the years from data taken by radioactive tracer [4.35], marker [4.30], and infrared isotope-shift [4.36] experiments. The more recent work by ROSENCHER et al. [4.37] on silicon oxidation in dry heavy oxygen ($^{18}O_2$) shows that the oxidant diffuses through the oxide layer previously grown in $^{16}O_2$ without any significant exchange with the existing light oxygen isotope and is incorporated only at the Si/SiO_2 interface.

In a similar experiment by ROCHET et al. [4.38], a thin (1.6 nm) layer of $Si^{18}O_2$ was formed on the top surface of a natural oxide film treated in dry $^{18}O_2$ for 8 h at 930°C. This may be due to network diffusivity from a surface layer [4.38], as MIKKELSON has recently qualitatively confirmed [4.39]. Although only a small perturbation to the overall reaction, this effect may become increasingly important in the early stages of oxidation.

Field-induced movement of ionic species has also been suggested, in addition to MOTT [4.40] and GROVE [4.21], by TILLER [4.4], who favoured an O_2^- influence, and LORA-TAMAYO et al. [4.23] proposed O^{2-}, and possibly O^-. Although the nature and charge of the oxidizing species remains topical in the sub-25 nm regime, the recent work of MODLIN and TILLER [4.11] who used a corona discharge, thereby eliminating any metal contacts with the oxide, has brought about a strong consensus of opinion in the field that neutral oxygen is at least the diffusion species during the oxidation of films thicker than 30 nm. In fact, these results appear to show that the previous observations of JORGENSON [4.30] discussed above are no longer valid.

HU [4.24] has considered chemisorption of the oxygen molecule followed by either direct oxidation, dissociation and then oxidation, or dissociation followed by ionization via tunnelling or thermionically emitted electrons. Strictly this is a more general application of the proposal of BLANC [4.28], who suggested an equilibrium between oxygen molecules and oxygen atoms at the oxidizing interface, and GHEZ and van der MEULEN [4.27], who similarly suggested simultaneous direct oxidation by oxygen molecules and indirect oxidation by oxygen atoms. Blanc in particular discussed the role of the broken silicon bonds at the reacting interface in determining the actual reaction rate during the early stages of oxidation. Under particular conditions, it was expected that the rate at which these bonds were made available for oxidation, most usually determined by the lattice temperature, could limit the rate of oxidation. Thence, the presence of ionizing radiation, or photons that are capable of producing a greater number of weakened bonds in the silicon, could somewhat influence and enhance the oxidation rate. The model of Blanc, however, does not predict the measured pressure dependence of the reaction, and as such cannot be taken in its present form as an acceptable alternative to other oxidation models. Nevertheless, the oxidation reaction has been regularly observed to be enhanced for Al, Si, and GaAs under optical illumination, and this introduces yet another angle from which to look at the basic reaction mechanism. This will be discussed later in this chapter.

Other ideas based upon simple extensions of the original Deal–Grove model include the incorporation of changes in diffusion resulting from stress [4.19, 20], or differences in diffusion, because of naturally occuring nm-sized channels, or much larger microchannels [4.17,18]. Compressive stress is built into the oxide at the interface with the silicon during the transformation from the regular crystal structure into the amorphous layer, since each unit of SiO_2 occupies around twice the original volume occupied by the silicon. Above temperatures of about 970ºC, viscous flow of the oxide may relax this stress, but upon cooling, further strain may develop because of the mismatch of the thermal coefficients of the materials. Any modifications to the structure, as well as the presence of any strain could not only affect the diffusion of the oxidant, but also indirectly the reaction mechanism itself.

FARGEIX et al.[4.20], and DOREMUS [4.19] found that the characteristics of the oxidation could be modelled by considering only stress – or strain – retarded diffusion of the oxidant. Rather than the diffusion of the oxidant being altered by strain or stress, IRENE [4.18] suggested similarly to REVESZ and EVANS [4.17] that diffusion through certain microchannels could account for at least the initial stage of rapid oxidation. Upon considering the influence of O_2^-, LU and CHENG [4.25] showed that an exponential distribution of the known charge content at the silicon/oxide interface into the oxide could account for the oxidation characteristics of both thick and thin films. Similarly, SCHAFER and LYON [4.26] have found that as the fixed charge becomes established when oxidation proceeds, it reduces the number of available Si bonds for the reaction by decreasing the hole population through band-bending.

Many of these models can be made to fit the growing wealth of experimental data for the early stages of the oxidation of silicon by using apparently realistic numbers for strain, charge content, and defect size. Several of them have also been supported by subsequent experimental evidence and observation, making it extremely difficult to preferentially accept any particular theory at this stage. It will probably be some time before experimental evidence is at hand to assist in the process of deciding which model eventually succeeds.

4.3 Metal Oxidation

In 1969, ASMUS and BAKER showed that the absorption of CO_2 radiation by thin metallic foils increased appreciably when the measurements were performed in an oxygen-rich environment [4.42]. This effect was correctly attributed to the laser-induced thermochemical reaction of oxygen with the surface atoms of the sample to form a thin oxide layer. In fact, this oxidation of metallic targets often became a problem for the users of high intensity CO_2 lasers for whom quality metal mirrors were essential. Throughout the 1970s several groups of Soviet scientists published the earliest studies of the mechanisms behind the formation of such films on various metals [4.43-46], although clearly this particular area of research did not blossom until the end of that decade and even then this work went largely unnoticed by the "new wave" of research. Interesting indeed is the quotation from a paper submitted at the beginning of 1972 by VEIKO et al. [4.45] which states "thus we may speak of localization of chemical reactions by use of the laser".

In this section we will discuss the use of lasers to encourage the localized formation of metallic oxides (Table 4.3). Although the great majority of these films are indeed insulating materials, some layers such as ZnO are actually semiconductors, and have only been included for the sake of completeness in this specific area of laser induced oxidation of metals.

Table 4.3. Laser-induced formation of oxides on various substrates (CW: continous wave, PD: pulsed)

COMPOUND	SUBSTRATE	AMBIENT	LASER	METHOD OF FORMATION	REF.
Al_2O_3	Si	$(CH_3)_6Al_2$ $+ N_2O$	KrF, ArF	PD CVD	4.135
Al_2O_3	Al	$(CH_3)_6Al_2$	ArF	CW CVD	4.136
Al oxide	Al	air	CO_2	CW heating & melting	4.46
CdO	Cd	air	Kr	CW heating	4.52,53
Cr_2O_3	Cr	O_2, air	Nd:glass	PD heating	4.59
Cr_2O_3	Cr	O_2,air	Nd:glass	PD heating	4.45
$\alpha-Cr_2O_3$ /$\gamma-Cr_2O_3$	Cr	air	CO_2	CW heating	4.159
Cu_2O	Cu	air	CO_2	CW heating	4.49,162
Cu_2O	Cu	air	CO_2	CW heating	4.44
Cu_2O /CuO	Cu	air	Kr	CW heating	4.52,161
In_2O_3 /InO	Quartz, InP,GaAs	$(CH_3)_3InP(CH_3)_3$ $+ O_2, H_2O$	ArF	PD CVD	4.137
Nb_2O_5 /NbO_2 /NbO	Nb	O_2	XeCl	PD heating	4.57
$Nb_2O_{5-\delta}$	Nb	O_2	CO_2	PD-plasma technique (LPPT)	4.138
SnO	Sn	air	Kr	CW heating	4.53,52
TiO_2	SiO_2	$TiCl_4 + CO_2$	CO_2	CW CVD	4.139
Ti_2O_3	$Ti_{50}Zr_{10}B_{40}$	O_2	Kr	CW heating	4.54
TiO /TiO_2 /TiO_2O_3	Ti	air	Nd:glass	PD heating	4.43
TiO_2	Ti	air	CO_2	CW heating	4.55
TiO_2 /Ti_2O_3	Ti films	air	Nd:YAG	PD heating	4.56
TeO_2	Te	air	Kr	CW heating	4.52,53
V oxides	V	air	CO_2	CW heating	4.48
V_2O_5	V	air	Kr	CW heating	4.52,53
WO_3	W	O_2,air	CO_2	CW heating /melting	4.46
ZnO	Zn	air	Kr	CW heating	4.52,53
ZnO	Si, Quartz	$(CH_3)_2Zn$ $+N_2O/NO_2$	Kr	CW heating	4.140
ZnO	Si, Al_2O_3, Au,Ti, Glass, GaAs	vacuum	CO_2	PD evaporation	4.140
ZnO	ZnO	vacuum	540nm	PD sputtering	4.132
CeO_2 HfO_2 Sc_2O_3 SnO_2 Ti_2O_3 Y_2O_3 ZrO_2		(O_2)	CO_2	CW (thermal) evaporation PD (flash) evaporation	4.142

149

Compound Semiconductors

Ga_2O_3 GaAs $/As_2O_3$		O_2	Ruby	PD melting	4.118
Ga_2O_3 $/As_2O_3$	GaAs	O_2	Ar	CW enhanced	4.87,119
Ga_2O_3 $/As_2O_3$	GaAs	O_2	Ruby	PD melting	4.73
Ga_2O_3 $/As_2O_3$	GaAs	O_2	Ruby	PD melting	4.143
In_2O_3 $/(InPO_4)$	InP	N_2O	ArF	PD CVD	4.133
Various	GaAs, InP,GaP	Air	Ar	CW enhanced	4.89

The most commonly applied laser method in metallic oxide formation is that of localized laser heating. This is conceptually the most simple and straightforward technique. It can be initiated by either continuous-wave (CW) or pulsed lasers, although in general the heating cycle is most controllable. with CW beams, while melting can be most easily achieved with pulsed radiation [4.47]. The laser wavelength must be chosen such that sufficient absorption takes place. Heating occurs when the absorbed energy is transferred to collective atomic vibrations, and in the presence of oxygen the reaction generally proceeds as would furnace oxidation. In this case, however, the chemistry is confined to those areas of the substrate elevated to sufficiently high temperature. With a knowledge of the activation energy the heating and cycle time can be adjusted to activate the reaction on the scale of micrometers.

Controlled heating of the substrate within an O_2 rich environment offers unique opportunities to study not only the early stages of the oxidation reaction but also the structural evolution of the oxide layers themselves. Such has been the drive behind the investigation of CW CO_2 laser oxidation of vanadium [4.48] and copper [4.49-51]. For example, it has been shown by URSU and coworkers for the case of copper oxidation by laser

[4.50] that there is generally good agreement between experimentally inferred values of oxide layer thickness in the early stages of growth, and the theoretical predictions of the Cabrera-Mott cubic growth rate law in the 100–250°C range

$$dx/dt = (d_1/x^2) \exp(-T_1/T) \tag{4.13}$$

and of the Wagner relationship for layers in the 50–100 nm range

$$dx/dt = d_0 \exp(-T_0/T) \tag{4.14}$$

where dx/dt is the rate of change of oxide thickness x with time t, d_0., T_0. The quantities d_1, T_1 are thermodiffusive constants, and T is temperature. Importantly, from an industrial point of view, the laser grown copper oxides adhere poorly to the metal, and often crack into small plates when scratched.

One notable point in this case was that only stoichiometric Cu_2O was observed for the thinner films. Small quantities of CuO were present only in thicker layers. Similarly, it was deduced by optical absorbance measurements that Cu_2O was formed when Cu was irradiated by CW krypton laser radiation [4.51]. However, more recent work by the same group [4.52] reports that as well as the usual red and green coloured zones associated with Cu_2O, a black external halo is observed. The fact that this feature is present before the usual Cu_2O is formed led to the postulation that CuO is indeed formed prior to Cu_2O, and that it is most certainly a necessary precursor to the rapid formation of Cu_2O.

Vanadium oxides, which exhibit semiconducting properties and have important applications as catalysts, come in many different and exotic phases. The evolution of each of the phases and the role of each in the initial stages of the reaction can be identified and studied under these novel laser-initiated process conditions [4.48].

Continuous-wave krypton laser irradiation of other metals in
air or O_2 has also led to the controlled formation of SnO, ZnO,
CdO, and TeO_2 [4.52,53]. By taking into account the thickness
dependence of the reflectivity and the heat of formation of the
reaction it has been concluded that the rapid oxidation rates ($\simeq$
10^4 nm/s for CdO) would only be attainable if the melting point
of the metal was reached. Similar irradiation of $Ti_{50}Zr_{10}Be_{40}$
resulted in the formation of amorphous films of Ti_2O_3 more than
100 nm thick with no evidence of any of the crystalline phases,
such as TiO_2, which is often the dominant structure formed in
conventional thermally oxidized Ti and Ti alloys [4.54]. Titanium oxidation by pulsed laser heating was first reported by
AKIMOV et al [4.43] who produced TiO_2 at the sample surface and
various lower oxides deeper into the Ti when irradiating the Ti
sheet with millisecond pulses from Nd:Glass laser. CW lasers
have also been used to oxidise Ti to form a TiO_2 [4.55]. Recent
work by THUILLARD and von ALLMEN [4.56] has produced novel
oxide structures on thin vacuum deposited Ti films on various
substrates. The group used a Nd:YAG laser, and found that
oxygen content and structure of the insulators produced was
strongly dependent upon the incident fluence. Additionally, the
electrical resistance of the films could be reversibly and continuously varied by up to 10 orders of magnitude.

Pulsed laser heating in the solid phase has also been applied
to form titanium oxides, using millisecond Nd:glass pulses in
air [4.43]. Shorter pulses from an excimer laser have produced
layers of NbO, NbO_2 and $Nb_2O_{5-\delta}$, characteristic of high temperature "thermal" oxides [4.57], in contrast to those formed by
Laser Pulse Plasma Chemistry (LPPC) [4.58]. The oxidation of
chromium to Cr_2O_3 in air using Nd:glass laser pulses has potential applications in mask making [4.45,59]. In fact, it has been
shown by METEV et al.[4.59] that such laser activated oxidation
permits the direct generation of patterns without the requirement of intermediate photolithographic processing more usually
followed in conventional microelectronic fabrication. Basi-

cally, an optical image of a mask is formed on the chromium surface by projection using 50 ns pulses to selectively oxidize the material. This has already been shown in Fig.3.10. Reductions of between 3x and 10x were used, and since the beam intensity on the actual mask itself was only 100 Wcm^{-2} emulsion masks could be used. Theoretical and experimental studies showed that irradiation with one pulse of 50 ns resulted in the formation of a 0.5-0.7 nm thick oxide layer on the 200 nm thick

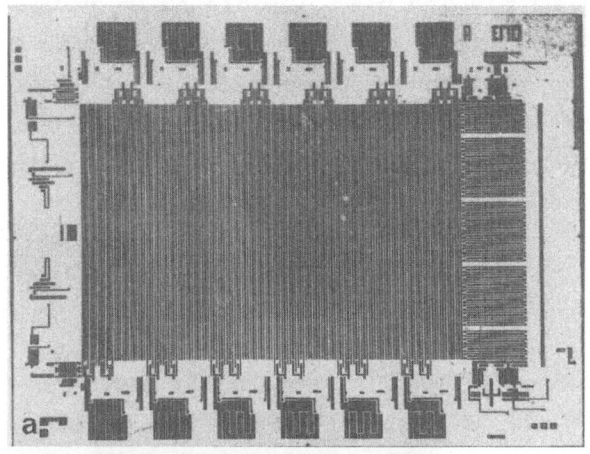

Fig.4.3. A photolithographic mask obtained by laser-induced oxidation (the darker regions) of a thin Cr film (the lighter coloured zones). The upper micrograph (a) shows an overall view of the mask, while the lower picture is a greatly magnified view of a small area in the lower section [4.59]

153

Cr film. Although this layer was found to be thick enough to permit selective etching of the unirradiated (i.e., masked) zones, up to five successive laser pulses were usually used to fabricate slightly thicker oxide films. Figure 4.3 shows a mask obtained by this photo-oxidation technique [4.59]. Etching was performed in solutions of $K_3Fe(CN)_6$, NaOH, H_2O. The total dimensions of the projected pattern were 9 x 7 mm^2 while the smallest feature resolvable was 3 μm, closely corresponding to the diffraction limit of the wavelength used. Visible or UV radiation should clearly further reduce the feature size available using this technique. The obvious advantages of using laser radiation for this application include the fact that if coherent radiation is used then the contrast of the image is more easily preserved.

4.4 Silicon Oxidation

The successful application of silicon in large scale device integration is due not only to the tailorability of its semiconducting properties, but also to the ease with which it can be effectively passivated by its natural oxide. Although various forms of silicon oxide exist, the most technologically important is silicon dioxide, SiO_2. It is an excellent insulator (resistivity $>10^{16}$ Ωcm), can withstand very large electric fields (dielectric breakdown strength $>$ 10 MV/cm), it passivates the silicon surface (surface states reduced to $<10^{11}$ cm^{-2}), and is chemically very stable.

Many methods of SiO_2 preparation are commonly available [4.60] and the physical and chemical properties of the differently prepared films can vary considerably. Direct thermal oxidation of the Si surface at elevated temperatures ($\simeq$ 1000°C) in an oxygen rich environment is the most commonly followed preparation technique. From the point of view of device performance, it has been found that such thermally grown layers are reproducibly quite stable and contain a very low level of interface state charge, mobile ionic charge and fixed charge [4.61]. However, there is an anticipated need for smaller geometry integrated circuits, in particular for very large scale integration (VLSI) applications, and an essential requirement for the fabrication of such devices is the minimization of high temperature processing techniques. It is known for example, that high temperatures can induce some degree of wafer warpage, and promote the generation of stacking faults and other defects, as well as encourage significant redistribution of dopants. Therefore, it has become increasingly desirable in recent years to form good quality SiO_2 films at lower substrate temperatures, or with a sizeable reduction in high temperature processing times.

It is worthwhile bearing in mind that although SiO_2 is currently the most widely used insulating material in the semiconductor industry, alternative dielectric materials are also the subject of intense research and are already finding limited application. These include, amongst others, silicon nitride, silicon oxynitride, and aluminium oxide. Several approaches to laser-assisted formation of these materials will also be described in the following sections.

4.4.1 Alternative Processing Techniques

Traditionally silicon dioxide layers have been prepared at temperatures between 800° and 1200°C in RF-heated quartz tube furnaces containing an oxygen-rich gas mixture. The oxidizing species diffuses across any intermediate pre-existing oxide layer and reacts at the silicon interface layer with the available silicon atoms. The gas mixtures can be pure O_2, O_2 in N_2, O_2 in HCl, O_2 in H_2; so-called "wet" oxygen where water vapour is present, or simply water vapour. In recent years, more general chlorine- and fluorine-containing gaseous mixtures have been used to prepare thin oxide layers. However, these processes usually require elevated substrate temperatures to encourage the reaction to proceed at an acceptably rapid rate, whilst maintaining dielectric integrity and the other desirable insulating properties.

The trend towards low temperature deposition of insulators started as far back as the 1960s. At this time, chemical vapour deposition (CVD) methods consisting of the pyrolytic decomposition of silane (SiH_4) in an oxygen environment were successfully introduced into the silicon integrated circuit industry to form silicon oxide protective coatings for the devices. Although, in general, these films can be grown at much faster rates than thermal oxides, their slightly poorer electrical properties have restricted applications mainly to the role of masking or protective overlayers. It has been found that an increase in electrical performance can usually only be obtained by increasing the deposition temperature. The CVD technique is discussed in more detail in Chap.6, where the development of laser–induced CVD (LCVD) is also described.

Reactive sputtering, in which Si atoms are ejected from a high purity source by energetic ion bombardment and consequently react with an oxygen rich atmosphere to form a thin layer of insulating SiO_2, has all but been replaced by RF sputtering from a silica target [4.62,63]. Under ideal conditions, RF-sputtered SiO_2 films can prepared that more closely resemble those grown thermally than other types of deposited films formed at the same temperature [4.64]. However close similarity has only been achieved by raising the substrate temperature, as with CVD films. Furthermore, layers prepared in this manner are usually subject to uniformity problems, while the substrate is prone to radiation damage, although these particular difficulties are continually being reduced [4.65].

Microwave plasma technology has also been investigated as a means of oxide formation in VLSI production [4.66,67]. In this case oxygen forms a reactive species that must be transported to the substrate to oxidize the Si. However, similar damage production and impurity incorporation threatens here too, to restrict potential applications. Progress in this area includes the idea of photon enhancement of the reaction [4.68]. Interest continues to develop in the area of high pressure oxidation, first demonstrated around 1960 [4.69]. More recently these high pressure techniques have yielded oxides whose dielectric strength and fixed charge content are as good as, or better than, those grown at one atmosphere [4.70].

4.4.2 Laser-induced Growth

Although CW lasers can be used to induce melting of the silicon surface, their general application in recent years has been in areas where controlled heating of the material is desired (Table 4.4). As in the case of metal oxidation previously described, CW lasers have been used to promote the growth of silicon oxides. The first report of laser induced surface oxidation of silicon in the sense of actual film formation, rather than oxygen contamination by incorporation during laser annealing and melting (Chap.5), was based on the application of a cw argon laser which operates in the blue/green region of the spectrum [4.71]. Since then many other groups have followed a similar method of controlled heating to produce oxides on silicon [4.72-78].

An attraction of the technique is that it is closely related to the traditional method of furnace oxidation, but enables localized rather than large area growth of silicon dioxide. As in furnace oxidation, the silicon is brought up to a predetermined temperature in a controllable fashion in the presence of oxygen where the usual first order chemical reaction occurs. However, furnace oxidation is performed under thermal equilibrium conditions, whereas, as pointed out earlier, CW laser heating and therefore laser oxidation is a highly transient thermal process. So while the wafers are heated fully across the surface as well as throughout their entire depth during

Table 4.4. Laser-assisted methods for silicon oxide formation (CW: continuous wave, PD: pulsed)

OXYGEN SOURCE /SUBSTRATE	LASER	METHOD OF FORMATION	REF.
Implanted O/c-Si	Nd:glass	PD annealing	4.148
Air/a-Si	Ruby	PD annealing	4.149
O_2/a-Si	KrF	PD annealing	4.150
Air,O_2/c-Si	Nd:YAG (1064, 532 nm)	PD melting	4.152
O_2/c-Si	Nd:YAG (532 nm)	PD melting	4.118
O_2/c-Si	XeCl	PD melting	4.104
O_2,CO_2/c-Si	Ruby	PD melting	4.143
O_2/c-Si	Ruby	PD heating	4.155
Air,O_2/c-Si	Nd:YAG (266 nm)	Amorphization	4.156,157
O_2,H_2O/c-Si	Ar	CW enhancement	4.71
O_2/c-Si	Ar	CW enhancement	4.72,77
O_2,wet O_2/c-Si	Ar	furnace heating & CW enhancment	4.73
O_2/c-Si	Kr,Ar	furnace heating & CW enhancment	4.75
O_2/c-Si	Kr,Ar	furnace heating & CW enhancment	4.76
O_2/c-Si	KrF	PD enhancement	4.97
N_2O/c-Si	ArF	PD enhancement	4.105
O_2/c-Si	Ar+CO_2	CW heating	4.78
Air,O_2/c-Si	CO_2	CW heating	4.72
O_2+NF_3/c-Si	ArF	Photo-oxidation	4.158
O_2/c-Si	ArF	Photo-oxidation	4.160
O Plasma/c-Si	CO_2	PD heating	4.68
SiH_4+N_2O/c-Si	ArF	PD LCVD	4.144,145
SiH_4+N_2O/c-Si	Kr	CW LCVD	4.146
SiH_4+O_2/c-Si	KrF	PD LCVD	4.138
Air+SiO	ArF	PD oxidation	4.91
Air + SiO_x	Ar	CW heating	4.166
$Si(OR)_x(OH)_{4-x}$ /c-Si	Ar	CW curing	4.131

furnace heating, in most forms of laser heating only the surface region immediately under the laser beam is subjected to the necessarily high temperatures, while the underlying layers remain relatively cool.

The early work on laser-grown oxide layers at Stanford using a tightly focused argon laser beam in O_2, and in steam [4.71], produced oxides of several hundred Angstroms at estimated growth rates of more than 0.3 nm/s. These were reportedly much larger than the corresponding thermal oxidation rate at the temperatures induced. Subsequent work in this regime found similar effects [4.72-76], leading to proposals that there was some "photon enhanced" effect in the reaction. The presence of the visible radiation during the reaction, it was suggested, could lead to increased ionization [4.71], an increase of available Si bonds for the reaction [4.73,76,77], or an assisted transfer of excited electrons into the pre-existing oxide layer to create more reacting species [4.75].

In order to isolate the possible photonic contributions to the reaction, it is important to know as accurately as possible the induced rise in the lattice temperature due to the de-excitation of the photogenerated carriers. Fortunately this problem can be approached through the use of well established theories of CW laser heating of silicon as discussed earlier in Sect.2.5. The general predictions of this theory have already been checked by independent groups against limited data for laser-induced melting of silicon [4.78,80]. For the laser oxidation work, the temperature rise induced by the beam was computed using (2.37) previously derived in Chap.2,

$$T = 99 + (T_0 - 99) \exp\left[\frac{NP(1 - R_0)}{2\sqrt{\pi}r(T_0 - 99)}\right]. \qquad (4.15)$$

In most applications, the samples are preheated to several hundred degrees Celsius. For high $P(1-R_0)/r$ ratios, preheating to 400°C actually adds as much as 1000 degrees to the eventual temperature attained as can be seen by studying Fig.2.18. In fact, preheating gives the added advantage of reducing the occurrence of slip dislocations on the surface (Fig.4.4) which often form as a result of large thermal gradients induced across the sample surface.

<u>Fig.4.4</u>. Slip dislocations formed on the c-Si surface as a result of steep localized thermal gradients induced by a CW laser beam [4.47]

Table 4.5 reviews the existing data for laser enhanced oxidation of Si by CW argon laser. The application of the formula to the known experimental conditions (column 2 of Table 4.5) gives a reasonable indication of the increased lattice temperature (1st column of Table 4.5). The original oxidation rate observed at T_0 (A) can then be compared with the increased rate due to laser heating (B) and the actual oxidation rate observed (C) (see column 3 in Table 4.5), to obtain the enhancement over the expected thermal oxidation rate (column 4 in Table 4.5). The enhanced rates are relatively small when the laser is used as the secondary energy source for oxidation, and are greatest when the laser beam supplies most of the energy for the reaction.

In the latter case, however, because of basic experimental uncertainty in the laser parameters, and the sensitivity of the oxidation rate to lattice temperature, the absolute degree of photonic enhancement is less accurately known. For example,

160

Table 4.5. A summary of cw Argon laser-induced oxidation rates for Si using published experimental details (references) and recently established theory (see text for complete details)

LATTICE TEMPERATURE before/after irradiation [°C]	LASER BEAM PARAMETERS [W/mm]	OXIDATION RATE (A)/(B)/(C) [10 nm/min]	PHOTON ENHANCEMENT [%]	REF.
900/910	1.1	2.5/2.9/3.2	10	4.76
900/923	2.4	2.25/3.1/3.3	7	4.75
880/902	2.4	1.85/2.4/2.8	17	4.75
810/831	2.4	0.75/0.95/1.05	11	4.75
770/790	2.4	0.36/0.49/0.56	14	4.75
820/841	2.4	0.65/1.06/1.29	22	4.73
400/866	6.4	10^{-5}/2.3/11	480	4.77
400/977	7.8	10^{-5}/9.8/22	220	4.77

even by assuming that the theoretical model used is precisely correct, there is a problem in determining the beam energy and radius to an accuracy better than a few percent. As shown in Fig.4.5, an uncertainty of 5% (or 50 degrees) in temperature around 1000°C results in a 3-fold increase in the predicted thermal oxidation rate due to laser induced lattice heating [4.82]. (The errors shown here also apply more generally to all areas associated with modelling growth rates induced by laser processing and they highlight the severe difficulties involved in obtaining an accurate knowledge of the induced temperature. Progress is currently being made in optical pyrometric systems for directly measuring these high temperatures with the required accuracy, although when the optical and thermal properties of the sample change as the film is growing, this will add an extra degree of complexity to the situation).

In recognition of this problem, a novel experimental approach has recently been reported that utilizes a naturally

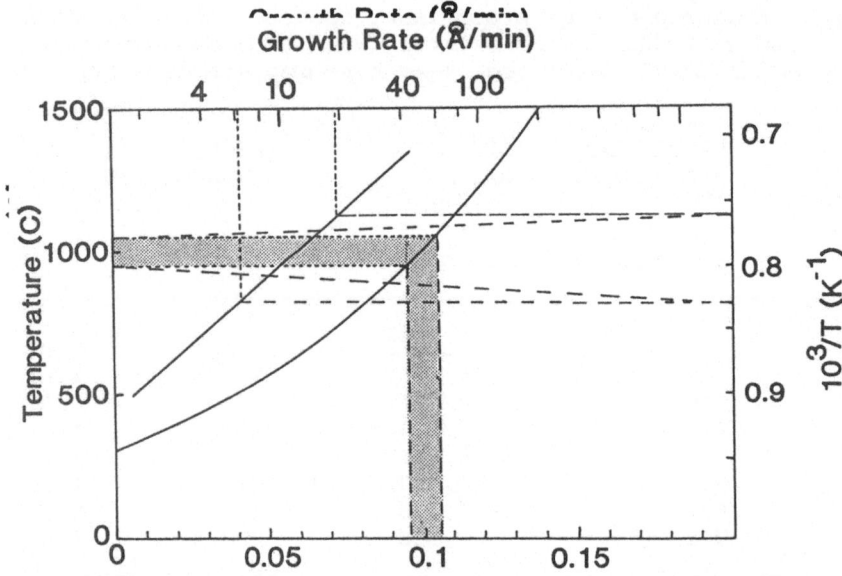

Fig.4.5. Plot showing how an uncertainty of ± 5% in the knowledge of the laser beam parameters leads to an factor of 3 uncertainty in the rate of oxide growth. The curve represents the theoretical temperature increase (assumed here to be 100% accurate for illustrative reasons) as a function of beam parameters, and the straight line is the thermal growth rate of SiO_2 as a function of reciprocal temperature. The necessary temperatures on each scale are correlated by dashed lines. Thus a $P(1-R)/r$ ratio gives an estimate of temperature through the curve, which is translated into inverse units and leads to a value for the growth rate [4.81]

occurring experimental amplification process to confirm the photonic effect [4.83]. The growing film changes the optical reflectance of the film/substrate combination because of the multiple interference effects. In doing so it increases the amount of power entering the material, and thence the induced temperature, resulting in a faster thermal oxidation rate.

If two similar wavelengths (488 and 514 nm) are used to oxidize the Si, initially set up so that the same power is absorbed at the start of the reaction, then any small changes in the induced reaction will cause the oxide to grow faster with one of the wavelengths. The faster growing film will cause the reflectivity to fall faster than the slower growing film, and thus produce a higher temperature increase and faster oxidation rate. In this way, small initial differences in the growth rate are amplified and after some time the thickness difference between the films can be easily measured.

It would be expected that since the 488 nm radiation is more strongly absorbed than that at 514 nm, a slightly higher temperature would be induced, and thus a faster thermal oxidation rate. This would be assisted by the fact that the reflectivity of the sample will decrease faster for the shorter-wavelength radiation due to the characteristics of the multiple interference. Fig.4.6 shows the findings of BOYD and MICHELI [4.83]. The reaction induced by the longer-wavelength radiation appears to consistently bring about a faster film growth, contrary to purely thermal expectations.

Having established that laser enhanced oxidation is thermally dominated, and photonic enhancement although consistently observed, is clearly only a second-order effect. Indeed there appear to be regimes where thermal effects completely overshadow any background photonic contribution. This does not, however, mask the potential importance of the photonic effect

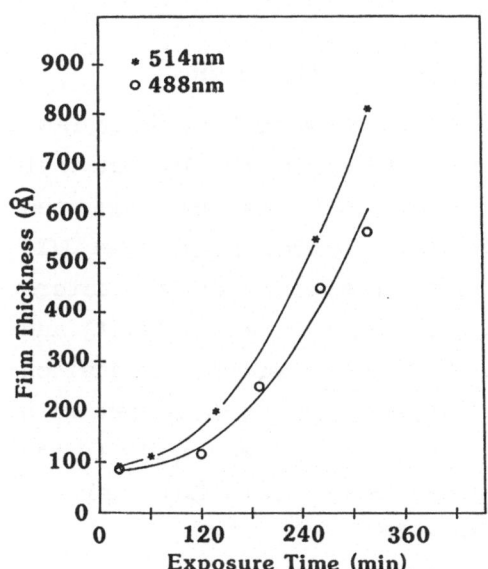

Fig.4.6. Oxide thickness as a function of exposure time for irradiation wavelengths of 488 and 514 nm [4.83]

under the appropriate conditions. For example, by taking full advantage of the photonic effect, the lattice temperature can be reduced appreciably whilst still achieving oxidation rates comparable to those presently used in VLSI fabrication, and this has important advantages for future device manufacture as discussed earlier.

A further complication in determining the kinetics controlling the photon–enhanced oxidation process lies in the basic lack of understanding and general controversy regarding the mechanism of conventional thermal oxidation as reviewed in Sect.4.2. So far, the theories of photonic enhancement have tended to favour modifications of the Blanc model [4.28], even though this model is known to be inconsistent with pressure-dependent data. However, more general models of GHEZ and VAN DER MEULEN [4.27], and HU [4.24] which encompass the Blanc model only as a special case, may provide alternative starting points for future analysis.

The laser oxidation theories have also been based on the premise that the reaction is rate–limited by the availability of Si bonds. In the next section, the various photonic enhancement models for laser induced oxidation will be reviewed.

4.4.3 Mechanisms of Photonically Enhanced Oxidation

There are now more than 20 publications showing that laser radiation not only induces oxidation of various materials, but that it also enhances the reaction somewhat. This has been suggested not only for silicon, but also for GaAs [4.85–88] and other III–V binary, ternary and quaternary compound semiconductors [4.89], InP [4.90] and SiO [4.91], as well as metals [4.51] and the not unrelated reactions of laser–enhanced nitridation [4.92] and etching [4.93] and the reaction of hydrogen with graphite [4.94]. Whilst some of the mechanisms may apply to one or more of the reactions, we shall concentrate here only on photonically enhanced oxidation of c–Si.

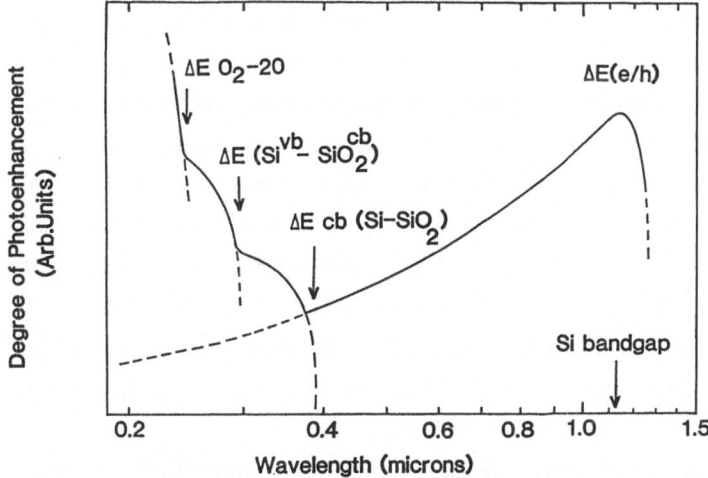

Fig.4.7. Spectral summary of the various mechanisms reported for the photonic enhancement of the c-Si oxidation reaction. Assuming the validity of each process, an estimate of the relative contribution of each has been made for a constant incident photon flux [4.95]

The term Photonic Enhancement (PE) used here embraces all reported effects in the light induced reaction that cannot be attributed to simple thermal mechanisms. Hence as in most non-thermal reactions the specific characteristics of the individual quanta of radiation are extremely important. Fig.4.7 is a brief summary of most of the proposed PE mechanisms for silicon oxidation reported in the literature, in terms of spectral response. An attempt has been made to estimate the relative contributions of the various effects for the constant incident beam power, and the reader is warned that not only is the relative importance of each open to debate, but in several cases, so are some of the enhancement effects. These are discussed below. It is important to recognize at this stage, that the photon flux directed at the c-Si surface during the enhancement will inevitably also induce some thermal effects, since none of the reactions reported to date are 100% quantum efficient. Therefore, one must also bear in mind that in almost every situation, there will be a background oxidation enhancement due to an increase in sample temperature.

The first report of light enhanced oxidation of silicon was by OREN and GHANDI [4.96], who used UV radiation from a mercury lamp during a conventional thermal oxidation reaction. They attributed the observed increase in oxidation rate to the photogeneration of electrons which are then liberated into the oxide leaving the c-Si in a highly charged state. They proposed a possible energy threshold to be that required by electrons to make a transition from the valence band of the silicon into the conduction band of the oxide, i.e. 4.25 eV. This corresponds to a wavelength of 292 nm; one of the most prominent lines of their mercury arc lamp was near 245 nm.

A similar proposition was made by SCHAFER and LYON [4.73], who applied low power laser radiation to a Si sample which was already undergoing thermal oxidation in a furnace at the usual elevated temperatures. Owing to the level of irradiation, however, allowances had to be made for the increase in temperature, and only small PE rates were extracted, typically of the order of 20-50%. It was subsequently noted that there was a threshold to the enhancement around 3-3.5 eV, which was attributed to the photo-excitation of conduction band electrons from the silicon into the conduction band of the oxide. Therefore, the threshold in this case was different from that of Oren and Ghandi by the silicon bandgap, i.e. (4.25 - 1.1 ≃ 3.2 eV). Coincidentally, this is remarkably close to the direct bandgap of c-Si, as shown in Chap.2. Although this may be important in terms of increased absorption, and therefore a rise in carrier population and a more elevated surface temperature, it does not eliminate the possibility of photonic rather than thermal enhancement of the reaction at this wavelength arising from the mechanism described.

The majority of the present models are based principally on the laser photogeneration of conduction electrons in the Si from the valence band, a process that occurs at photon energies near to and above 1.1 eV. In chemical terms, this may be likened to antibonding. GIBBONS [4.71] proposed photo-ionization. The enhancement is then related to the new equilibrum of photo-generated carriers above that established by the usual intrinsic thermal generation. In the rate limited oxidation regime, for the growth of thin films às discussed earlier, the reaction may be limited by the availability of unsaturated silicon bonds. However, in the more general sense, very similar models can be produced for cases where the actual electron (or hole) concentration is crucial to any chemical or physical process near the surface.

YOUNG and TILLER [4.76] have shown that the enhancement is directly proportional to the number of photons incident during the reaction rather than the power density, in the blue/green range of the Ar laser. Therefore, in order to obtain the same enhancement rates for different wavelengths, the photon fluxes can be adjusted so they are equal for each wavelength used. Thus, less power is required by lower frequency radiation to achieve the same enhancement as higher frequencies. This means that if one applies the 647 nm line from the Kr laser, and the same absorbed photon flux from the 350 nm line of the Ar laser to induce oxidation, the same enhancements should be obtained even if the power of the former was only half that of the latter. Alternatively, if one preheats the Si mildly to around 400°C, where the Si oxidation is extremely slow, and repeats the experiment with Nd:YAG laser radiation (at 1064 nm) and with the 350 nm Ar line, almost a factor of three of beam powers would show the same effects. Of course, additional heat will be supplied by the relaxation and recombination of the photogenerated carriers and this may often mask the photonic effect on the reaction. Owing to the wavelength dependence of the silicon absorption coefficient, the shorter wavelength will tend to heat up the silicon much more efficiently. YOUNG and TILLER also reported a second photonic effect, that a bigger enhancement is also found with <100> orientation Si than with <111> material. Because of the different density of atoms in each crystal plane this seems to be contrary to what may be expected for a purely thermal process.

The final category of enhancement modes included in this diagram is the photodissociation of molecular oxygen close to the silicon substrate. The bond energy of O_2 is known to be close to 5.1 eV, corresponding to a wavelength of 240 nm. Any photons of energy greater than this can crack the molecule into two extremely reactive oxygen atoms, which may either react instantly with the silicon or the oxide layer, or alternatively with other oxygen molecules to form ozone (O_3), which is also a much more powerful oxidant than its allotropic cousin.

A recent study by FIORI [4.97] has resulted in the postulation of laser-induced bond rearrangements at the silicon surface in the presence of adsorbed oxygen and radiation of the appropriate wavelength. It was found that although a monolayer of adsorbed oxygen did not form any structures resembling silicon oxide after thermal treatment, if the oxygen rich surface was subjected to ultraviolet radiation at 248 nm ($h\nu \simeq 5eV$) and not, incidentally, 305 nm radiation at the same intensity, then a disordered structure of silicon oxide was formed. Subsequent thermal annealing at temperatures up to 949 K eventually produced a stable amorphous SiO_2 layer. The effect of the laser radiation, therefore, was a crucial step in the oxide formation. The wavelength threshold for the process is interestingly close to that previously found in the laser oxidation experiments SCHAFER and LYON [4.75] described above. Whether or not this mechanism of laser-induced incorporation of weakly bonded oxygen species into the silicon surface is the controlling factor in these photon controlled enhancement reactions remains to be confirmed by more experimentation. Nevertheless, it appears that a new mechanism for oxide formation specifically at wavelengths around 248 nm has been revealed. Indeed, layers up to 20 nm thick have been grown in 300 min using low UV laser powers ($6 \ W/cm^2$), and it has been suggested that this is by no means an upper limit to the thicknesses available with this technique [4.97].

At this stage, it is relatively straightforward to incorporate the effects of visible radiation on the reaction to most of the oxidation models requiring the presence of electrons, or holes, or partially ionized silicon. It is not so easy to accommodate models involving strain-effected diffusion as the limiting factor in the growth of thin layers, even though this may have a strong influence on the overall oxidation rate. Under these conditions, the photogeneration of carriers or additional oxidizing species may enhance the reaction if they diffuse much faster than those prevalent before the photoexcitation.

Finally, we discuss the work of BLUM et al. [4.91] showing that evaporated silicon monoxide (SiO) films can be readily photo-oxidized in air using a pulsed excimer laser. As will be discussed in the following section in detail, the peak position of the Si-O absorption band near 1075 cm^{-1} can occasionally provide a reasonably good indication of the oxygen content of these oxide layers. Fig.4.8a shows the position of this band as a function of the number of laser shots of 110 J/cm^2 at 193 nm incident on a 400 nm thick SiO film. In conjunction with helium backscattering analysis, this indicates that the SiO film gradually oxidized to SiO_2. In Fig.4.8b the fluence dependence of the reaction is plotted. The films were irradiated at different fluences in such a way that they were always subjected to the same total fluence of 110 J/cm^2, and the diagram shows that the films tend to oxidize more readily when irradiated by higher fluences. Since similar pulses can melt c-Si at fluences around 500 mJ/cm^2 [4.98] it may be expected that each 110 mJ/cm^2 could increase the surface temperature by approximately 300 K. The authors found, however, that even by irradiating films preheated to 340°C, (and thus probably bringing them up to temperatures of the order of 640°C) there was no detectable influence on the laser-initiated process. Three possible mechanisms have been considered

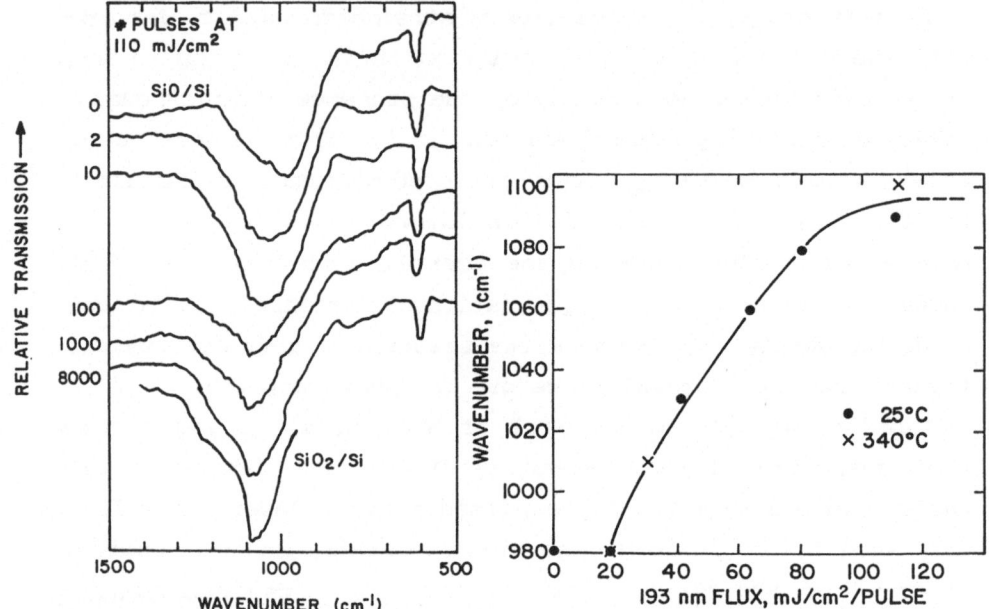

Fig.4.8. (a) Shape of the 1075 cm^{-1} absorption band of a 400 nm photo-oxidized SiO film as a function of the number of laser pulses at 110 mJ/cm^2. The curves have been displaced vertically for clarity [4.91]. (b) The effect of incident fluence on the conversion of SiO to SiO$_2$; see text for essential detail [4.91]

$$SiO^* + SiO \rightarrow Si + SiO_2 \qquad (4.16)$$

$$SiO^* + O_2 \rightarrow SiO_2 + O$$
$$O + SiO \rightarrow SiO_2 \qquad (4.17)$$

$$O_2 + h\nu \rightarrow 2O^{\cdot}$$

$$O^{\cdot} + SiO \rightarrow SiO_2 \qquad (4.18)$$

Mechanisms (4.17,18) are apparently consistent with the observed difficulty in photo-oxidation of thick rather than thin films [4.91]. However, further experiments are required to ascertain which of these particular mechanisms has the controlling influence on the reaction. Since the SiO films were initially deposited at room temperature and the laser processing performed at temperatures estimated to be less than 300°C, this

170

may be yet another possible method for producing SiO_2 layers without raising the substrate temperature substantially.

4.4.4 Rapid Thermal Oxidation

CO_2 lasers have also been used to controllably oxidize silicon since the early 1980's [4.72,74,99]. Since 1985, there has also been a faster growing interest in the use of of intense incoherent lamp furnaces to also grow oxides (and nitrides) rapidly. Both methods involve heating rates that are typically several miliseconds to several seconds, rather than minutes, which are more representative of traditional RF-heated furnaces. The clearest advantage of these techniques is that the heating up and cooling down times, which can often be longer than the actual film growth time, become a relatively insignificant consideration. Consequently, the high temperature processing times for oxidation can be dramatically minimized, thereby helping to eliminate warpage, contamination, and undesirable diffusion. The significant difference between heating silicon with infrared rather than visible radiation is that absorption occurs via the excitation of free carriers to higher energy states within the conduction band, rather than directly exciting carriers across the bandgap. Hence the number of free carriers will be essentially the same as that determined by the lattice temperature, and therefore no photonic enhancement to the reaction is expected.

The CO_2 laser would not normally be considered to heat up silicon in such a controlled way as to be useful for localized pyrolytic reactions. In fact, the absorption of 10 μm radiation in silicon is extremely weak, being typically only 1-2 cm^{-1} [4.78]. However, an important practical application of the argon laser required that the silicon substrate be preheated by some means to around 300-400°C in order to eliminate the formation of slip dislocations discussed earlier. Preheating the silicon to 400°C increases the absorption coefficient at 10 μm by nearly an order of magnitude and this is adequate to allow mod-

171

erate CW beam powers to heat crystalline silicon to typical oxidation temperatures. In fact the silicon can be seen to glow red when temperatures around 700°C are reached. By gradually increasing the incident CW laser beam power this surface glow changes to orange, then yellow, and brilliant yellow once the crystal has melted. But before the melting point is reached, the c-Si begins to buckle and warp when irradiated for pro- longed periods due to the once again excessive thermal gradi- ents. Even below this level slip lines begin to appear once again and therefore a processing window must be defined [4.99]. The beam powers must be sufficient to induce useful oxidation rates but not so high as to promote deleterious side effects to the silicon wafer. One distinct advantage of using CO_2 lasers for techniques such as this is the usual Gaussian-type tempera- ture profile induced by the heating beam. Here, films can now be grown at a range of temperatures simultaneously, under oth- erwise precisely identical growth conditions. A considerable difficulty may be introduced by having to know the precise tem- perature very accurately, but for fundamental studies of the relative temperature- or thickness-related characteristics, this may be a useful tool.

Under the most ideal conditions, oxides as thick as 180nm have been grown, at a rate of 2nm/min. The same thickness was found on both sides of the c-Si wafer indicating that the tem- perature was the same on both front and back surfaces and therefore uniform throughout the depth of the sample. This also indicated that there was no photon enhanced reaction on the directly irradiated surface, as originally expected for this wavelength. As is well known, SiO_2 is optically transparent in the visible, and its presence was visually established by a series of coloured bands due to the interference of light from the front and back surfaces of the layer. In fact, the colour of the interference ring is directly related to the thickness of the layer.

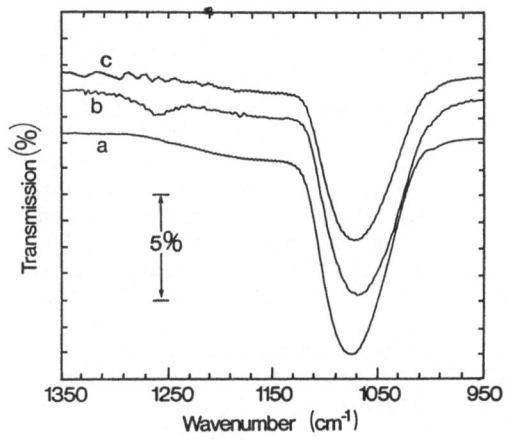

Fig.4.9. Infrared transmission spectra in the 950 – 1350 cm^{-1} range of three differently prepared oxide layers on silicon. The absorption band is that of the Si-O stretching vibration. The layers were grown (a) by furnace in H_2/O_2 at 950°C; (b) by furnace in HCl/O_2 at 850°C, and (c) by a CW CO_2 laser in O_2

Infrared spectrometric analysis revealed that the films were SiO_2 rather than SiO or Si_2O_3 as indicated by the presence of the dominant Si-O stretching vibration near 1075 cm^{-1} rather than at 1000 cm^{-1} for SiO or 1040 cm^{-1} for Si_2O_3 [4.74]. A typical spectrum is shown in Fig.4.9 and is compared with the spectra of other oxide layers grown by conventional techniques. The precise position and width of this band has often been used by others in the past to indicate not only the oxygen content of the films, but also as a means of obtaining information on the strain, porosity, and density on films formed mostly by deposition techniques, or by other novel methods where oxygen content of the films may be non-uniform [4.100-102]. However, it is known from a great wealth of surface analytical techniques, that thermally grown oxide on on c-Si becomes stoichiometric after only a few monolayers. Therefore, by studying the characteristics of this absorption, one may obtain information on features within the layer other than oxygen content.

The peak position of this 1075 cm^{-1} band as a function of film thickness is shown in Fig.4.10a for both laser-grown oxides (LO) and furnace oxidized layers (FO) [4.103]. While the peak position of the FO settles around 1070-1072 cm^{-1} for the thicker films studied, it tends towards 1078 cm^{-1} for LO. Both sets of data appear to have the same overall behaviour. For

173

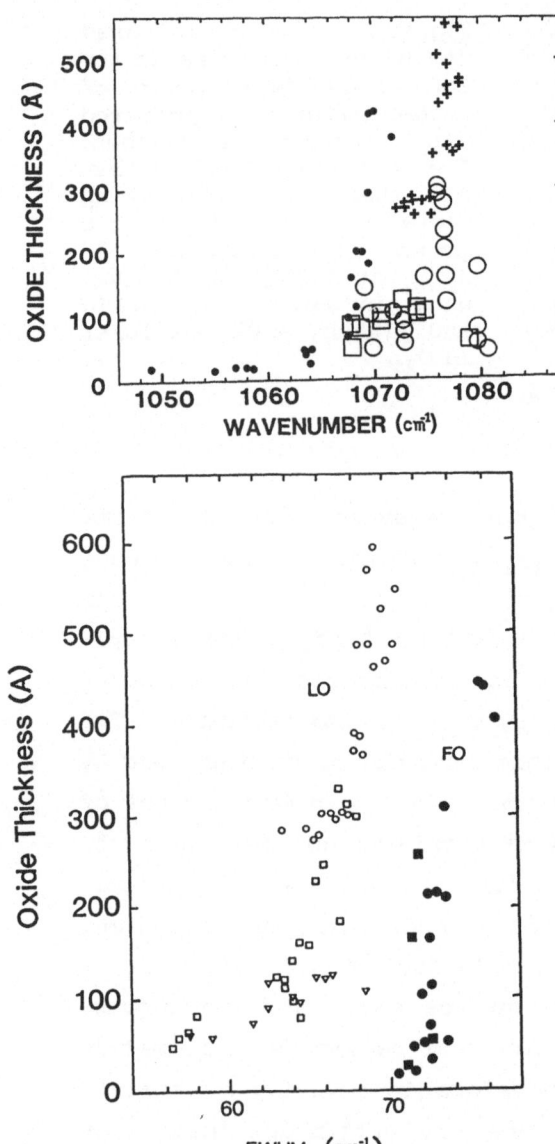

Fig.4.10. (a) Precise position of the Si-O absorption band minimum near 1075 cm^{-1} as a function of film thickness. The filled circles represent conventional furnace grown layers, each datum being from a different sample, while the remaining +, O and symbols are from a variety of oxides grown on three separate c-Si substrates using the CO_2 laser [4.103]. (b) Spectral width of the absorption peak around 1075 cm^{-1} of various CW CO_2-laser grown silicon oxides (LO), and some furnace grown layers (FO), as a function of film thickness. Each symbol for the LO represents a different c-Si substrate (although a range of films of different thickness were grown on each), while (o) and (•) represent a range of films grown by two different furnace oxidation processes, at different temperature [4.103]

thinner oxides, below about 15 nm, the peak position of the band shifts strongly towards lower wavenumbers. For reasons outlined in the paragraph above and elsewhere [4.103], this has been interpreted in terms of strain, emanating from the interface with the silicon.

174

The spectral width of this band is consistently narrower for the laser-grown layers than for the furnace produced films. This is detailed in Fig.4.10b. It is important to note that the laser-grown layers were formed at a variety of temperatures across the surface of the irradiated sample, while the conventionally prepared films were produced only at two different temperatures. Although two sets of data are distinctly separated by some 6 cm^{-1}, the general trend of a gradually increasing spectral width with film thickness is very similar in both cases, except for the thinnest layers. Therefore it would seem that for films thicker than about 10-15nm, this behaviour is independent of formation temperature. These results can be compared with the recent data obtained using a UV melting technique [4.104], which show a 50% increase in this width.

Although the oxides grown by solid-phase heating are known to be amorphous, the changes in width of this band have been attributed to changes in the natural distribution of bonding. In other words, an increase in the width of the 1075 cm^{-1} band is due to an increase in the spread of Si-O bond angles. This is supported by the fact that the width of this band is consistently wider for CVD films than for thermally grown films, and it is well known that the CVD layers are generally much more porous and contain a greater density of voids and defects than do furnace grown layers. This implies that the laser grown films are somehow less disordered than the furnace grown layers. Further studies, involving Ar laser oxidation, and perhaps annealing steps, may shed some light on the situation. Whatever the interpretation, the CO_2 laser grown films are structurally very similar to conventionally formed furnace oxides, unlike those produced by the UV laser melting method.

The changes in the spectral width of both sets of films also exhibit a narrowing for thicknesses less than 15 nm. As with the data for the shift in peak position, this has been attributed to strain from the c-Si interface as the oxide structure attempts to grow from the ordered crystalline base. Although

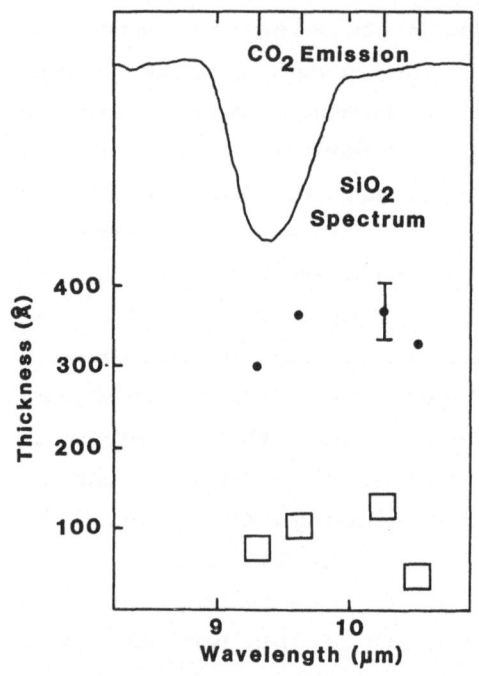

Fig.4.11. The four main emission lines of the carbon dioxide laser are shown at the top of this figure, together with the Si-O stretching vibrational band. Also detailed is the thickness of various silicon dioxide films grown by the laser using each of the four lines shown, for two different exposure times

some 10-15 nm is required to grow before the IR features are no longer affected by these effects, the extent of the actual strained region will be considerably less than 10 nm.

Since SiO_2 absorbs more strongly at 9.3 μm than at, e.g., 10.6 μm it may be expected that the growth rate of the oxide layers would be influenced by which of these heating wavelengths are used. Fig.4.11 shows the four main emission lines of the CO_2 laser and the thicknesses of various films grown by each of the lines for two different irradiation times. Within experimental error no wavelength dependence is apparent. However, as mentioned above, these films are grown under conditions where the temperature is evenly distributed throughout the c-Si. Therefore, any increase in absorption at the front surface of the sample will not cause an increase in average temperature, since the laser energy absorbed within the sample is still the same.

By comparison, TOKUYAMA et al. [4.68] have observed a wavelength dependence to the oxidation rate by using the CO_2 laser to enhance film deposition in a microwave plasma. In this case, the oxide thickness of 100 nm was irradiated with the various lines from the CO_2 laser and whilst the extra oxide growth was found to be 2.5 nm for the 9.3 μm wavelength, an increase of only 1.5 nm was found for the 10.6 μm irradiated sample despite the beam powers in the latter case being larger. Clearly, Toku-yama et al. were not operating in the regime of uniform heating throughout the sample.

Upon re-examination of the spectral details of Fig.4.10, one may ask why the CO_2 laser grown oxides exhibit different IR spectra from furnace grown layers. There are several differences in the processing environments. Firstly, although the carrier populations may be in principle identical, the electronic excitation pathways between the systems is quite different. The electron density on each silicon atom may be crucial to the formation of specific bonding arrangements. Thus, instead of the usual sp^3 hybridization producing mainly sigma bonds with the oxygen, d-type orbitals may initiate pi-bonding with the p-type bonds of the oxygen. However, there is not yet any experimental evidence for such a mechanism.

Another important difference in the growth environments is that in furnace oxidation the gas-phase molecules may be considered to be in thermal equilibrum with the furnace and sample. On the contrary, with the degree of localized laser heating, it is not so clear that the impinging oxygen molecules are at the same temperature as the hot silicon. Consequently, in one case we have hot molecules reacting with the hot silicon, and in the other, cold molecules reacting with the heated substrate. This may not only give rise to different oxide structures, but different reaction mechanisms. Detailed studies are underway to examine this reaction possibility.

Despite the variations in peak position and width of the 1075 cm^{-1} band, its absorbance, A, can be related to the ellipso-metrically determined thickness of the oxide film. Absorbance is defined by

$$A = \log_{10}(I_0/I) = 0.434 \ln(I_0/I) \tag{4.19}$$

where I/I_0 is the fraction of the incident light transmitted by the film. Knowing Beer's law defined earlier in Chap.2, gives

$$A = 0.434\alpha(\text{apparent}) \cdot x \tag{4.20}$$

where x is the total film thickness, and $\alpha(\text{apparent})$ is the effective absorption coefficient. Knowing that absorption in the front and back surface oxides contributes to A, while the ellipsometrically measured thickness only refers to one side, this equation must be scaled by a factor of two to give

$$\alpha(\text{apparent}) = A/0.868x \tag{4.21}$$

Plotting A versus x, $\alpha(\text{apparent})$ is found to be $3.4 \pm 0.1 \cdot 10^4$ cm^{-1}. Thus the thickness of a grown oxide layer, $x(\text{grown})$, can be calculated from the infrared absorption at the band minimum using the equation [4.6]

$$x[nm] = x(\text{grown}) + x(\text{native}) = 3.42 \cdot 10^2 \, A \tag{4.22}$$

where $x(\text{native})$ is the native oxide thickness ($\simeq 1.5$ nm, see Fig.4.1b) and A is measured when c–Si with only a native oxide layer is placed in the reference beam path of the ratio–recording spectrophotometer.

The absorption of 10 μm radiation can be increased in silicon not only by preheating, or by increasing the doping concentration, but, more novelly, by introducing by optical means, more free carriers into the material. In fact the new technique of dual beam heating (and oxidation) of silicon using the argon laser to promote the free carriers and the CO_2 laser to provide the power for the temperature rise, has been successfully used to grow oxide layers thicker than 200 nm [4.78]. This method offers the dual advantage of the diffraction limited resolution of the argon laser (0.5 μm) and simultaneously the enormous availability of power from the CO_2 laser, which is one of the most efficient laser systems available.

In a similarly novel vein, pulsed laser irradiation of crystalline silicon in an atmosphere of N_2O rather than the more usual O_2 has recently been shown to produce insulating oxide layers similar to those grown in pure O_2 [4.105]. The growth rate of these films has been estimated to be 0.3 nm/min, which is comparable to the rates achieved by the more usual solid-phase laser-induced reactions, and this suggests that the surface does not melt under the irradiation conditions used in this experiment.

The development of rapid lamp oxidation has attracted considerable attention in the semiconductor industry in recent years. For potential applications that demand large area processing, rather than diffraction limited spatial resolution, an array of intense, incoherent, essentially white-light sources, can serve as an excellent heating source for many materials. Several different types of arc-lamp, as well as tungsten filament systems have been used by several groups [4.106-117]. Heating rates associated with this technique are typically 500-1000 K/s, so that a 10nm thick oxide layer can be grown in less than 2 min. By introducing different gases into the sample chamber during heating, different types of oxide, as well as nitrides and other layers, can be formed.

Recent analysis of the oxidation rates achieved by these methods have suggested that the reaction may actually be photonically enhanced in some way, similar to that previously reported oxidation of the c-Si by visible and UV laser radiation. PONPON et al. [4.110] have recently found a possible enhancement in their measured oxidation rate, while CHAN TUNG et al. [4.111] reported a rapid initial oxidation rate of up to 1.1 nm/s during the first 5 seconds of the reaction and just over 0.3 nm/s during the next 5-60 s at 1250°c. Activation energies of 0.9 eV [4.111] and 1.2 eV [4.112] have been extracted from the growth data for the rapid oxidation regime, while 1.4 eV has been calculated for the slower post-5s oxidation [4.111]. These observations do not agree with extrapolations of the available data for conventional oxidation. For comparison, MASSOUD et al. [4.116] found a 3.2 eV activation energy for their conventional thermal oxidation at temperatures above 1000°C. Together with the laser-induced oxidation experiments reported since the beginning of the decade and discussed above, there appears to be strong evidence for previously undiscovered oxidation mechanisms. Further work is obviously required to establish whether this RTP procedure is purely thermal in nature. Whether or not any photonic mechanisms are present in these reactions, it is clear that this "Optical Furnace" is becoming a most important piece of semiconductor processing apparatus.

4.5 Oxidation of Compound Semiconductors

We have seen that since oxidation of c-Si requires one of the highest processing temperatures on the standard microelectronic production line, the laser has attracted a lot of attention in the search for alternative lower temperature oxide formation methods. A further advantage of many of these techniques, is that they can be easily applied to materials where it is much more difficult to form stable insulating and passivating layers. For example, compound semiconductors such as GaAs normally dissociate at the usually moderate temperatures required to deposit conventional CVD insulators. This is one of the main reasons why many technologies based upon binary compounds have been slow to evolve.

Early experiments involving pulsed laser processing of compound semiconductors actually induced melting with segregation of the constituent atoms. Therefore, even though a mixture of hexagonal Ga_2O_3 and cubic As_2O_3 was formed when the GaAs substrates were irradiated in O_2, the surface layer down to 50 nm was found to be Ga rich and As deficient. Similar experiments using almost identical irradiation conditions, showed that the laser intensity had to be nearly three times that required to melt the GaAs before any oxygen could be incorporated into the lattice [4.118]. Indeed between the melting threshold E_{th}, and $3E_{th}$ no oxides could be formed. It has since been shown that the native oxide must be evaporated at this higher intensity before significant new oxide formation can occur by oxygen penetration through the melted region [4.119]. Table 4.3 has already summarized the various approaches to laser production of compound-oxide layers on compound semiconductors.

The formation of oxides on GaAs without surface melting is clearly a much more attractive proposition. PETRO et al.[4.87] have shown that low intensity (<3 Wcm^{-2}) Ar laser irradiation of cleaved GaAs substrates increased the sticking coefficient of O_2 by a factor of 1000. Heating effects and oxygen excitation were ruled out as possible causes of the enhancement in favour of electronic excitations of the GaAs surface [4.87]. More recently, the same group [4.120] further postulated that in addition to this excitation the oxidation reaction may proceed via a precursor absorption of a loosely bound oxygen molecule, followed by break-up of the molecule into reactive species. Evidence for this model indicated that in the dark the uptake of oxygen by the GaAs was strongly temperature dependent, while under laser illumination the chemisorption was totally independent of thermal activation. Coverages of up to 2 monolayers were found, with no evidence of saturation.

BERMUDEZ [4.88] has shown that irradiation with similar intensity incoherent radiation also produced photoenhanced oxidation of GaAs in the presence of O_2, but not in CO and H_2O atmospheres. Since the latter molecules do not chemisorb, this suggests that the intramolecular bonding in the oxidizing species must be weakened for the reaction to occur. The bonding in the adsorbate is weakened by a charge transfer from the bulk material, and with the photoenhanced breaking of Ga-As bonds, oxidation thus proceeds. Enhancement factors of up to 20 have been observed in the times required to form the first monolayer of oxide, while the rate of photoenhancement does not strongly depend upon the type of excess carriers present in the material. Whilst experiments with NO indicated a much smaller sticking coefficient and subsequent oxide growth than with O_2, BERTRAND [4.84] found that oxidation of GaAs irradiated by 25.4 and 18.5 nm lamp wavelengths occurred in the presence of N_2O. In this case the As rich surfaces oxidized much more slowly, and saturation occurred once all the adsorbed oxygen became incorporated to form the oxide.

SIEJKA et al. [4.85] have similarly observed preferential oxidation of Ga, in addition to some loss of As during 193 nm laser irradiation of GaAs ^{18}O-enriched native oxides in an $^{16}O_2$ atmosphere. Additionally, very efficient exchange of oxygen was observed after 1000 laser pulses, with the total oxygen content of the films being conserved. Since thermal heating was believed to be essentially non-existent, the authors appear to have found a new mechanism of laser assisted formation of highly mobile oxygen species in GaAs.

Another independent study using CW argon laser radiation also found optically enhanced oxidation rates for GaAs, and for many other III-V and ternary compounds [4.89]. As a result it was found that after initial rapid oxidation, film growth proceeded in proportion to the logarithm of illumination time when intensities of about 1 kW/cm^2 were used. Semiconductors containing Ga oxidized rather easily during the initial stage of irradiation, while those containing As were oxidized more easily than the others either after a certain amount of film growth, or at temperatures near 300°C. The average temperature increase induced by the laser was estimated to be only a few tens of degrees. However the growth rate was seen to be enhanced as if the temperature had been increased by about 50°C, as shown in Fig.4.12. Since the logarithmic oxidation behaviour did not change under laser illumination, it was suggested that the enhancement was most likely due to accelerated transport of the elements necessary for oxidation.

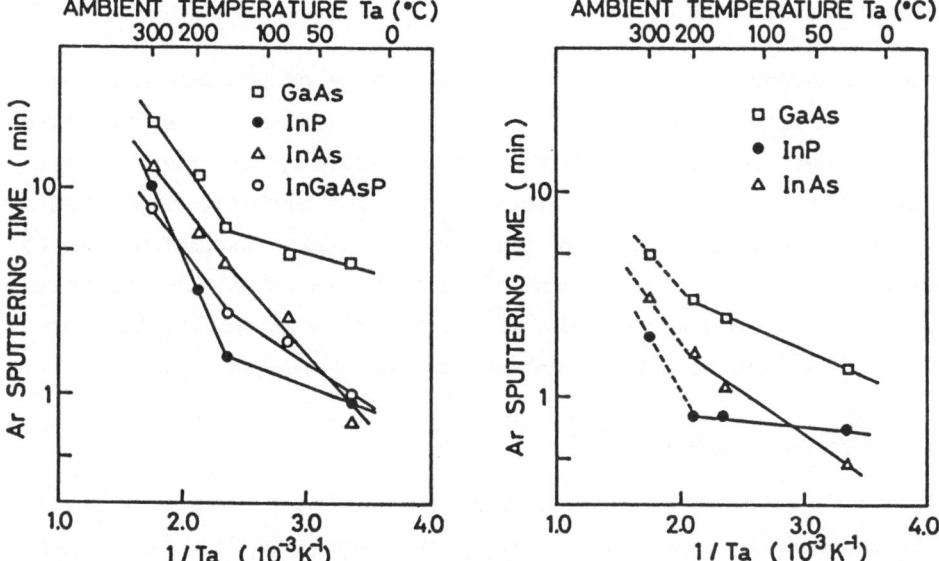

Fig.4.12. Arrhenius plots of oxide thicknesses, determined by sputtering in (a) irradiated, and (b) unirradiated portion of the sample. The kinks in the curves of (b) can be seen to be displaced by approximately 50°C from the curves in (a) [4.89]

Finally, the oxidation of InP has also been achieved using an ArF laser in an atmosphere of N_2O [4.90,105]. The process is most likely initiated by the photodissociation of the N_2O molecules, since it was found that for a given constant laser power the oxide thickness increased with increasing N_2O pressure. When the laser beam impinged directly on the sample surface during the film growth, the oxidation rate was enhanced by approximately 20%. In agreement with the photon enhanced Si oxidation models, and some of the related ideas connected with enhanced reaction rates also discussed throughout this book, it was suggested that the increased concentration of electron-hole pairs could assist in the breakage of the lattice bonding, thereby increasing the availability of atoms for the reaction. Films as thick as 16 nm have been grown using this method, and these appear to be amongst the thickest laser-grown oxides on compound semiconductors reported so far.

4.6 Nitridation

Just as oxygen can be reacted with a range of materials to produce surface oxide layers, nitrogen can, under the appropriate conditions, be encouraged to react similarly and produce nitride layers. These films not only exhibit usefully high values of microhardness, but on silicon are also characterized by extremely attractive insulating properties for VLSI devices. Here we review the methods of nitride growth, whilst in Chap.6 the mechanisms of nitride deposition are reviewed.

4.6.1 Thermal Nitridation

Very thin oxide layers are not perfectly impervious to impurity diffusion and it is well known that intense radiation can generate a large density of interface states in the oxide causing a degradation of the device performance. There are also frustrating yield problems associated with the growth of layers thinner than about 10 nm. Consequently, possible alternative forms of insulating layers are always being investigated. Thermally grown silicon nitride, as well as thermally nitrated SiO_2, presently appear to be good alternatives to the pure oxidation process, exhibiting those properties lacking thin oxide layers mentioned above. Some of the additional advantages of nitridation include a self-limiting growth which is therefore easily controllable and a high oxidation resistance. Furthermore the films produced are chemically very stable.

Silicon nitridation has been investigated in a number of ways, as summarized in Tables 4.6a and b. For example, ITO et al.[4.121], first reported silicon nitridation in purified ammonia (NH_3) or in N_2 at temperatures around 1200°C. CHANG et al.[4.122] subsequently reported the growth of similar layers in a nitrogen plasma at 600°C, while ITO et al [4.123] also performed plasma enhanced thermal nitridation in an RF generated ammonia plasma at temperatures around 1000° C. In the past few years, interest in this topic has increased considerably and

Table 4.6. (a) Summary of recent processes used to initiate nitridation: Non-laser methods. (b) Laser-assisted methods for nitridation (CW: continuous wave, PD: pulsed)

GAS	SUBSTRATE TEMPERATURE [$^{\circ}$C]	METHOD	REF.
Ne or NH_3	1200	thermal	4.121
N plasma	600	plasma	4.122
N plasma	114	plasma	4.124
NH_3	950–1230	thermal	4.114
NH_3 plasma	100	plasma	4.123,125
N_2/H_2 plasma	700–900	plasma	4.126

(a)

AMBIENT	LASER	METHOD OF FORMATION	REF.
Silicon Nitride			
NH_3 + SiH_4 + N_2	ArF	PD CVD	4.144
NH_3 + SiH_4 + Ar	ArF	PD CVD	4.136
NH_3 + SiH_4 /Si_2H_6 + He	ArF	PD LCVD	4.163
NH_3 + SiH_4 + Ar	CO_2	LCVD	4.165
NH_3 + SiH_4	CO_2	Reactive nucleation	4.134
N^+ implanation	Nd:glass	PD annealing	4.148
NH_3	ArF	PD nitridation	4.92
NH_3 + SiH_4	CO_2	Reactive nucleation	4.144
Si_3N_4	CO_2	CW/PD evaporation	4.142
Other nitrides			
PH_3	ArF	PD CVD	4.147
$Zr+N_2(g)$ /$N_2(liq)$	CW CO_2	nitridation	4.128,130
$Ti+N_2(g)$ /$N_2(liq)$	Nd:YAG	nitridation	4.129,130
Fe + $NH_3(liq)$	XeCl	nitridation	4.164

(b)

further work on thermal nitridation between 950° and 1230°C in
NH_3 [4.114], and on plasma nitridation in nitrogen [4.124], NH_3
[4.125], and N_2 and H_2 [4.126], have been reported. The work of
MOSLEHI and SARASWAT [4.114] involved the use of a lamp heating
system, and reported reliabilities of the thin nitride layers to
be far superior to SiO_2 and nitroxide because of reduced charge
trapping. Layers thinner than 4 nm showed high current leakage
due to the onset of quantum mechanical tunnelling, implying
that films thicker than this must be used in typical FET's, as
is the case with the oxide layers. Lasers have been little used
in this area, but probably the most exciting work is that of
SUGII et al. [4.92], who induced excimer laser enhanced nitridation of silicon. This will be discussed in more detail in the
next section.

4.6.2 Laser Nitridation

Using the 6.4 eV photons from an argon fluoride excimer laser,
SUGII et al. [4.92] produced thin nitride films on silicon surfaces with incident fluences of only 15 mJ/cm^2 in each pulse,
and in the presence of ammonia gas. In this instance the substrates were held at a pre-irradiation temperature of 400°C,
and it was calculated that the induced temperature increase
was of the order of 55°C for each pulse, while the repetition
rate of the laser was only 12 Hz. The conclusion drawn from
these figures was that the increase in thermal activity was not
responsible for the observed growth of the nitride layer.

Figure 4.13 shows the thickness of the films produced by
this method. It can be seen that there is a limiting process, as
observed with thermally produced processes, with the thickness
levelling off at about 2.4 nm, despite the increasing number of
shots. This saturated thickness is similar to that found when
c-Si is thermally nitrided at around 700°C, as shown by WU et
al. [4.127]. In Fig.4.14, the similar Auger electron spectra of
the laser grown films and those produced thermally at 1000°C

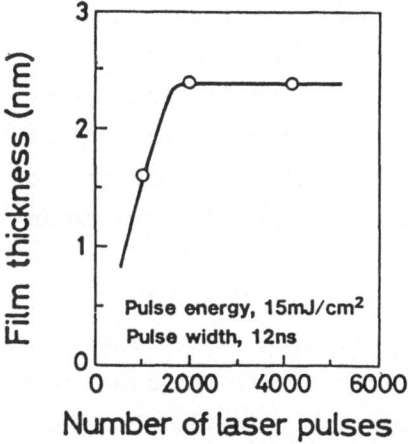

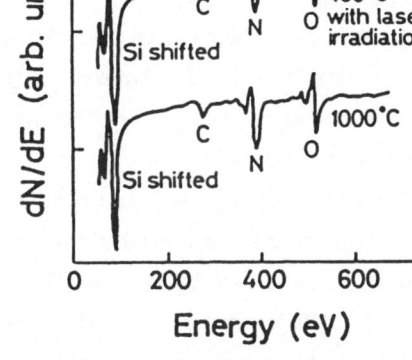

Fig.4.13. Thickness of silicon nitride film grown by excimer laser, as a function of number of pulses [4.92]

Fig.4.14. Auger electron spectrum of silicon nitride grown at 400°C by excimer laser, and at 1000°C in a furnace [4.92]

implies that at least the surface components of each film are virtually identical.

It is known that the optical absorption of ammonia is more than 100 $atm^{-1}cm^{-1}$ at this wavelength [4.92]. The controlling mechanism is the photogeneration of NH_2 radicals in the gas phase which subsequently react with the silicon surface. It is possible that, because the limitation on the layer thickness was due to the restricted diffusion of the nitridation species across the grown layer, thicker films could be grown with only a modest increase in the substrate temperature.

Similar work on laser induced nitridation, but of metallic surfaces, has also been reported recently by URSU et al. [4.128-130]. This group has shown that by irradiation of various metals, especially zirconium and titanium in air, technical nitrogen, and liquid nitrogen, using a CW CO_2 laser or TEA-CO_2 laser, a wide range of surface hardened nitride and oxynitride layers could be prepared. The TEA-CO_2 laser is thought to generate a plasma near the irradiated surface which assists in the production of the required species [4.129]. It has also been suggested that surface melting is an additional prerequisite for significant nitride formation in this regime. Plasma induced damage was minimized by using longer laser pulses and higher gas pressures. While it required several hundred pulses to grow a 1 μm thick nitride layer in gaseous N_2, only about 20 pulses were necessary to grow the same thickness of films in liquid nitrogen [4.130].

The process by which nitride films grow on Ti and Zr in air is strongly influenced by the presence of oxygen and the amount of heat liberated by its exothermic reaction with the substrate [4.128]. (This point is discussed more fully in Chap.7). Thus the nitride layer produced in this way is initiated at temperatures in excess of 1000°C, and is always sandwiched between the substrate and an oxide layer. Indeed, it was found in all of these experiments that the presence of trace amounts of oxygen greatly affected the precise reaction induced on the sample surface, as well as the composition, quality, and physical properties of the grown film.

4.7 Laser Curing

KRCHNAVEK et al. [4.131] have developed a laser direct writing technique for forming thin insulating layers of silicon dioxide on various substrates. The technique involves coating the substrate with organo-silicate film and thermally curing it with a scanning laser beam, which in this case was from a CW argon ion laser. A family of films, of the formula $Si(OR)_x(OH)_{4-x}$, initially undergoes an elimination reaction to expel the organic liquid, when irradiated. At increased temperatures the OH groups are driven off leaving a layer of SiO_2. Lines of continuously varying thickness can actually be drawn using this technique, by changing the translation speed and power density of the beam. In fact, linewidths as low as 1 μm have been obtained (Fig.4.15). Indeed, it is possible, since the quality of these layers is at least as good as that obtained in thermally cured

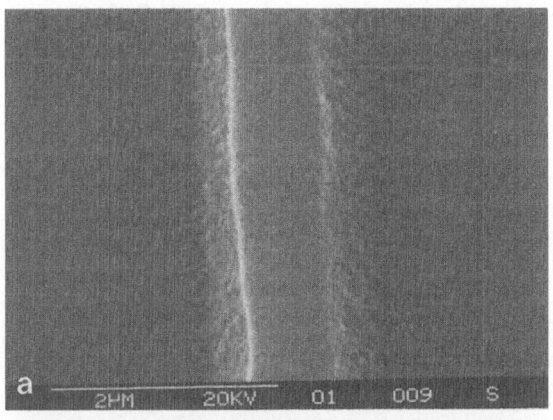

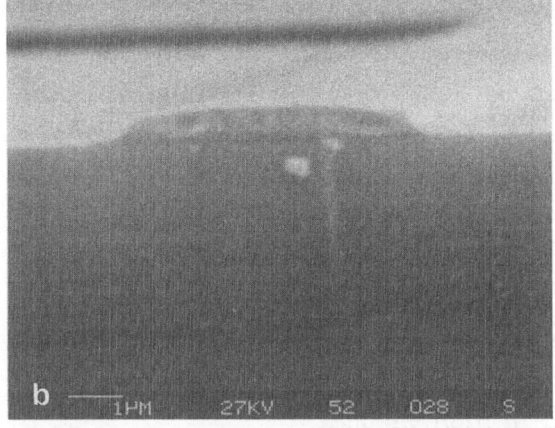

<u>**Fig.4.15**</u>. Scanning electron microscope pictures of a one micrometer wide line of silicon dioxide, drawn by a CW argon laser at an intensity of 4 MW/cm^2: (a) top view, (b) cross section [4.131]

films, that new device structures requiring a variable oxide thickness could be developed. Also, an ability to directly write such structures locally, or alternatively, to change the refractive index of particular oxide-based materials by only a few percent could have a dramatic impact on optical waveguide processing.

5. Passivation by Laser Annealing and Melting

Whilst Chap.4 deals with controlled passivation of surfaces in the solid phase, Chap.5 summarises the more extreme processing conditions that have been applied to the formation of oxides and nitrides. Two main regimes are described here. The first, laser annealing, involves oxygen or nitrogen implantation into the semiconductor, followed by laser induced melting and epitaxial regrowth of the disordered layer. An alternative method employs similar high intensity laser radiation to melt a thin layer of amorphous (or crystalline) material in the presence of e.g. oxygen. These novel non-equilibrium processing techniques have enabled film growth rates of more than mm/s to be achieved. This chapter also describes how surface cleaning can be performed by applying a succession of high intensity laser pulses to the material held in high vacuum.

5.1 Modes of Laser Annealing

In the materials processing industry, annealing has traditionally been a process by which solids, especially metals, are heated up to some temperature and then slowly cooled over a carefully regulated time period, so that any strain within the bulk can be relieved. In the semiconductor production lines, ion-implanted silicon, disrupted by the penetrating high energy ions, is similarly annealed in well controlled RF heated furnaces in order to recover the initial high degree of crystallinity of the material. Annealing times are typically from 30 minutes to several hours, depending on the degree of damage within the samples. The use of laser radiation to bring about a similar improvement in lattice reordering and the accompanying increase in conductivity associated with such structural rearrangements, has in fact been sporadically reported over the years since the late 1960s [5.1-6]. However, it was not until the more complete studies of several Soviet groups during 1974 and 1975 [5.7,8], and others later [5.9-13] that the phenomenon of laser annealing became a popular topic for investigation. (It is interesting to note that the term "annealing" was applied to such a process, since it is now known, as it was strongly suspected then, that extremely rapid thermal transients, sometimes of the order of 10^{15} K/s, and even substrate melting, are involved in the ordering mechanism). Since then, many hundreds of papers have been published on the subject, dealing not only with the practical aspects of the operation, but also the fundamental mechanisms behind the various well-defined processing regimes established.

Laser annealing can be conveniently classified into two categories, solid phase annealing, akin to the furnace annealing process traditionally used, and liquid phase annealing, where localized melting is induced at the surface of the material. Annealing in the solid state involves solid phase epitaxy, and the rate of regrowth from the damaged (amorphous) phase to the crystalline phase is primarily determined by the temperature at the reordering interface. Other factors such as the dopant concentration, dopant type, crystal orientation etc., also have some effect on the regrowth rate, and because the complete procedure involves nonequilibrium heating, it is not a simple task to quantify the the total operation. Nevertheless, models such as those presented in Chap.2, have been shown to be extremely useful and apparently extremely accurate in such analyses.

Since a prerequisite for annealing is that the sample must be heated to some elevated temperature, reactions other than simple atomic rearrangements can clearly be induced by laser beams. In fact, at times it becomes important to consider the chemical constituents within the immediate environment as undesirable chemical reactions could be initiated at these increased temperatures. In the previous chapter, the effect of gaseous oxygen on the surface during laser-induced solid phase heating of many types of materials was reviewed. In the present section, the presence of oxygen, in particular, during the transition from the amorphous phase to the crystalline phase, i.e. the annealing process, and from either phase to the liquid phase, i.e. the melting process, will be considered. The oxygen need not be in the gas phase, of course, and it can also be present as a result of deliberate implantation of oxygen ions. For example, there is a strong interest presently in the possibility of preparing buried oxide layers inside the surface of silicon, as a means of manufacturing radiation hardened microelectronic devices, and a wide variety of annealing conditions are currently being investigated in this area.

Even though CW lasers primarily induce solid phase recrystallization, they can also initiate liquid phase epitaxy by locally melting the amorphous layer. However, this is more usually the regime of pulsed lasers, which can deliver the energy necessary for such annealing in times of the order of the pulsewidth, i.e. nanoseconds, or even picoseconds. Under these conditions, regrowth rates can be 5 orders of magnitude faster than in the solid phase, and the redistribution of impurities in the irradiated layer is indicative of diffusion coefficients as much as 7 orders of magnitude larger than those characterizing solid phase diffusion. As pointed out in Chap.2, these temperature gradients and rapid atomic movements can give rise to a host of completely novel processing conditions never previously available, leading to, for example, the formation of supersaturated solids. The consequences of these extreme operating conditions in terms of the formation of oxide and nitride films on silicon will be discussed in this chapter. It is possible, with the current development of even larger ion implanters for potential application in the formation of buried oxide layers, that other forms of insulating and doped layers could eventually be similarly investigated.

5.2 Laser Annealing in Oxygen

The applications of silicon dioxide in the semiconductor industry are numerous. These layers are used to passivate, to provide electrical isolation between layers or neighbouring devices, to provide under controlled conditions an electrical path between elements, for isoplanarization, to make excellent microalloyed contacts of aluminium and silicon, as protective overlayers, as dielectrics in capacitive devices, and of course as a mask to provide blanket protection during ion implantation. More exotically, silicon dioxide has been considered as a possible diffusion source in self-aligning schemes, or in localized oxidation schemes that form the basis of a technology where device geometries can be significantly reduced, while oxide isolation techniques can provide near perfect isolation between pockets of crystalline silicon on an insulating substrate. This latter application is useful for the fabrication of microelectronic components that must be radiation hardened, and also in special high voltage applications.

Reference to Table 4.4 in the previous chapter will show that one of the most common means for inducing oxidation of silicon with lasers has been by initiating silicon phase changes in the presence of oxygen. Pulsed lasers have invariably been used for this purpose, at wavelengths from the UV (excimer lasers) through the visible to the infrared (CO_2 laser). In fact the role of atmospheric oxygen, present during laser annealing of silicon in air by pulsed lasers, has been questioned for some time [5.14]. Naturally, when it was first found that ultra-short laser pulses could be used to produce thin amorphous layers on single crystal silicon substrates irradiated in air [5.15,16] the influence of the oxygen in the ambient during the amorphization cycle was examined more closely. It was tentatively suggested that oxygen may actually be incorporated into the material during its liquid phase in such quantities that it hinders any form of crystallization. LIU et al. [5.17], however, reported that the presence of a wide range of different gases including oxygen during the process did not have any controlling effect on the crystal to a transition to the amorphous state. It has since been shown by several groups that this unusual phase change is a consequence of the unprecedentedly large thermal gradients induced near the surface, so that after melting, the material does not have sufficient time to recrystallize when it solidifies and the amorphous state of the liquid is frozen into the solid state. This work of LIU et al. was performed using picosecond pulses.

By comparison it was noticed by GARULLI et al. [5.18] who used nanosecond pulses that one of the disadvantages of pulsed laser annealing of ion implanted layers was the in-depth contamination by oxygen during the actual annealing process. These studies showed that a considerable amount of α-quartz (a crystalline form of silicon dioxide), was formed after amorphous silicon samples were annealed in air with Q-switched ruby laser pulses.

Similar experiments performed in pure oxygen environments using a CW laser to anneal previously amorphized layers also revealed the incorporation of oxygen atoms into substitutional lattice sites within the silicon [5.19]. In this case, it was shown that the structure of the oxide components was most likely β-crystobalite, another form of crystalline SiO_2. Fig.5.1 shows the Reflected High-Energy Electron Diffraction (RHEED) patterns that were obtained from amorphous silicon samples laser annealed in O_2 by a CW CO_2 laser. Also shown is the pattern obtained from identical material that had been laser annealed in vacuum. Clearly there are many more diffraction features on the O_2 annealed samples. The accompanying Tables 5.1,2 compare the interplanar spacings extracted from the measure-

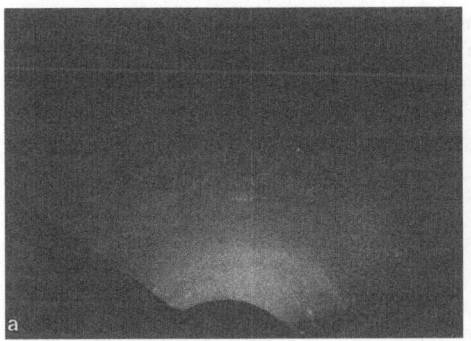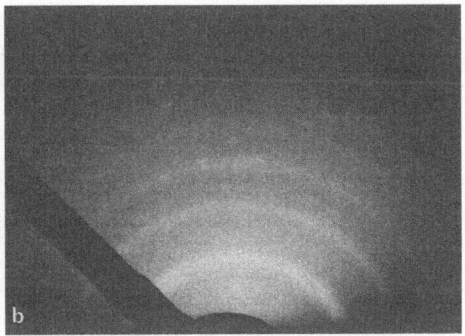

Fig.5.1. RHEED pattern obtained for silicon, pre-amorphised by ion-implantation, and annealed (a) in vacuum for 30s, (b) in oxygen for 30s, by a CW carbon dioxide laser [5.19])

Table 5.1. Comparison of the tabulated (ASTM) interplanar spacing (d_{hkl}, in nm) for poly-Si with those obtained from the RHEED pattern for a-Si CW laser annealed in vacuum (shown in Fig.5.1a)

RING RADIUS [cm]	d_{hkl} [nm]	d_{hkl} [ASTM,Si] [nm] (hkl)	REL.INT. [%]
1.25	0.394	–	–
1.6	0.308	0.314(111)	100
2.3	0.214	–	–
2.55	0.193	0.192(220)	60
3.0	0.164	0.164(331)	35
4.0	0.123	0.125(331)	13

ments with those tabulated for β–crystobalite, and it is evident that there are some contributions from this crystalline form of SiO_2.

These experiments indicate that the incorporation of oxygen into the reordering material is strongly dependent upon the temperature during the annealing cycle, on whether melting occurred, and on the availability of the oxygen itself during the process. Of course, one need not start off with amorphous material in order to incorporate oxygen into the solid during

195

Table 5.2. Values of interplanar spacing d_{hkl} in nm, calculated from the RHEED pattern of a-Si CW laser annealed in O_2 (shown in Fig.5.1b) are compared with the ASTM figures for poly-Si and β-crystobalite

RING RAD. [cm]	d_{hkl} [nm]	d_{hkl}(Si)	REL.INT. [%]	d_{hkl}(β-Cris) [nm]	REL.INT. [%]
1.15	0.429	–	–	0.415	100
1.65	0.299	3.14(111)	100	–	–
1.95	0.253	–	–	0.253	80
2.35	0.210	–	–	0.207	30
2.55	0.193	1.92(220)	60	–	–
2.9	0.167	1.64(331)	35	–	–
3.15	0.156	–	–	0.164	60
3.45	0.143	–	–	0.146	50
3.95	0.125	1.25(331)	13	0.127	30
4.5	0.110	–	–	0.121	30

irradiation. In other words, the annealing process is not an essential step in the mechanism of oxide formation. This has been shown by a large number of studies involving irradiation of single crystal silicon in oxygen rich environments in the solid phase (Chap.4) and the liquid phase [5.20-25]. Basically, all that is required in the first instance is that the material is heated up for a sufficient time that oxygen atoms can become incorporated into the silicon and react chemically with it.

It is perhaps rather surprising that appreciable oxygen can be incorporated into the silicon when 30 ns pulses are used to heat up (but not melt) the material [5.23]. CROS et al. have monitored the gradual growth of the first few monolayers of oxide on freshly prepared silicon surfaces heated by several 30 ns pulses of ruby laser radiation in a low pressure ($4 \cdot 10^{-4}$ Torr) oxygen environment. The pulses of 1.3 J/cm^2 apparently did not melt the surface of the sample, and after 27 pulses it was found that a saturated film thickness of 0.65 nm was grown. Subsequent irradiation with these pulses did not measurably increase this layer thickness. These films could not be grown when the gas phase oxygen was not present, ruling out the possible contribution of interstitial oxygen, always present in bulk microelectronic-grade silicon. However this still leaves the question as to how sufficiently large numbers of oxygen atoms at these low pressures strike the silicon surface and take part in the oxidation reaction when the silicon is irradiated for only 30 ns. It can be calculated, by assuming ideal gas law conditions, using

$$M_i = 3.513 \cdot 10^{22} \ P/\sqrt{mT} \qquad\qquad (5.1)$$

where m is the mass of the oxygen, and T the temperature, that the impingement rate M_i, of oxygen molecules on the sample surface at a pressure $P = 5 \cdot 10^{-4}$ Torr is approximately $1.5 \cdot 10^{17}$ atoms/s/cm^2 at room temperature. Since there are about $2.2 \cdot 10^{15}$ oxygen atoms/cm^2 present per nm thickness in SiO_2, this implies that at least 10 ms would be required to supply enough oxygen for each nm to grow, assuming a sticking coefficient of unity. The sample is probably heated for around 500 ns after each laser pulse, the total irradiation leading to a heating time of 1.5 μs. Clearly this is not sufficient time for any significant oxygen coverage to develop on the surface.

This anomaly can be explained by the reasonable assumption that the oxygen is adsorbed on to the material surface in adequate amounts before the irradiation takes place. This suggestion is also supported by the fact that layers much thicker than two monolayers cannot be produced by this method. Ways of producing thicker layers might involve increasing the total heating time by orders of magnitude or even inducing surface melting by increasing the incident beam intensity. The former approach has already been used with CW lasers, as discussed in Chap.4. Here, we will explore the possibility of processing with higher intensity pulses.

By applying (5.1) above to atmospheric pressure of oxygen, it can be estimated, using the same assumptions as before, that enough oxygen could be supplied from the gas phase to grow 0.1

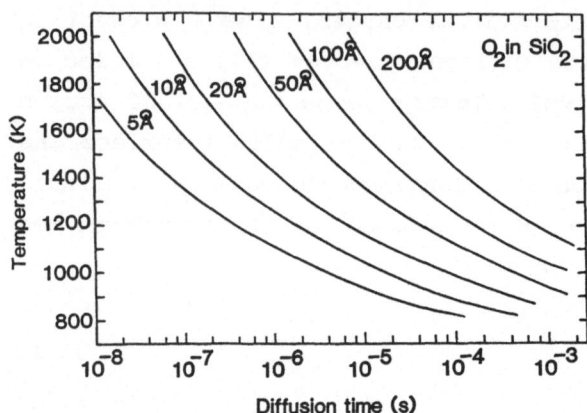

Fig.5.2. Time required by O_2 molecules to diffuse through various thicknesses of silica at a constant temperature [5.45]

nm of oxide every nanosecond. If this were possible, then growth rates of 1 cm/s should be achievable. However, the adsorbed oxygen must diffuse through any oxide layer already formed in order to meet the reactive silicon at the $Si-SiO_2$ interface. The diffusion of O_2 molecules through silica of various thicknesses is shown in Fig.5.2, using the expression

$$t = L^2/5.4 \cdot 10^{-4} \exp(-1.16q/kT) \tag{5.2}$$

where t is the diffusion time, L is the silica thickness, and 1.16 eV is the activation energy and $2.7 \cdot 10^{-4}$ the frequency factor in the usual Arrhenius diffusion characteristic [5.26]. It can be seen that even for temperatures above the c-Si melting point of 1685 K up to the melting point of SiO_2 ($\approx$2000 K) the process will be diffusion limited once layers of 5 nm or so are formed. This thickness is considered to be very much an overestimate, since such temperatures would probably not be sustained for extended periods of time in pulsed laser processing.

This suggests, therefore, that although the first few monolayers may grow extremely rapidly, any oxide grown will tend to hinder the necessary diffusion required for further oxida-

198

tion to occur during subsequent laser irradiation. Even surface melting may not be sufficient to induce the rapid growth of thick oxide films on silicon if temperatures approaching the oxide melting point are not achieved. An extrapolation of the known data [5.27] reveals that even at 2000 K, the dissolution rate of the solid SiO_2 into the liquid Si would be around 50 nm/s, so that an oxide film of 1 nm would take 20 ms to dissolve into the melt. This is also many orders of magnitude longer than the known duration of any pulsed laser induced liquid phase.

As will be shown in the following outline, extremely thick oxide layers have been grown by pulsed laser irradiation of Si in oxygen. In fact the growth rates discussed above have been achieved on several occasions. It is rapidly becoming clear, however, that although such astounding growth rates may be possible, the films grown are not of microelectronic device quality, and this seems to be a consequence of the extremely intense pulses that are usually applied to the Si, and the highly nonlinear and nonequilibrium kinetics that control the reactions.

Secondary Ion Mass Spectroscopy (SIMS) measurements of ^{18}O in crystalline silicon samples that were irradiated and melted in $^{18}O_2$ by intense 150 ns pulses showed that ^{18}O was introduced into the lattice with a maximum concentration of $7 \cdot 10^{20}$ cm^{-3} and a penetration depth of 1.4 μm [5.20]. Yet it was shown that the incorporation of oxygen was stopped by a 90.4 nm thick layer of SiO_2 on the sample surface.

SCHELL-SOROKIN and DEMUTH [5.28] have used UV photoelectron spectroscopy (UPS) and high resolution electron energy loss spectroscopy (EELS) to study the behaviour of chemisorbed oxygen at submonolayer coverages on laser annealed Si<111>, after irradiation by a 25 ns multimode ruby laser pulse in the fluence range 1.0-2.2 J/cm^2. They have in fact reported removal of surface contaminants by repeated irradiation, confirming the work of ZEHNER et al. [5.29] and others, which will be discussed later in Sect.5.3. Results indicate that under single irradiation events oxygen is not desorbed, but is transformed into clumps of SiO_2 scattered unevenly across the irradiated region. The exposure conditions used here are expected to bring about surface melting, with a melt duration of approximately 100 ns [5.30]. The authors suggest that this is so short that normal equilibrium conditions are not reached and competing silicon recrystallization may interfere with the oxide formation, although it is possible that the inhomogeneous laser beam profile affects the uniformity of the reaction. The spectroscopic information also indicates that the oxides formed are not typical of ideally bonded tetrahedral silicon dioxide, but exhibit features that may be related to known network and point defects in vitreous and crystalline SiO_2.

Short pulses (15 ns) of UV radiation which also induced surface melting have similarly been used to grow thin oxide layers on silicon. However these short pulses also create such fast melting and resolidification conditions that the liquid silicon does not recrystallize but actually freezes in an amorphous state [5.24,25]. Nevertheless there still appears to be sufficient time for enough oxygen to become incorporated during melting to produce oxide layers approaching 30 nm in thickness. More recent work in this regime [5.22] has produced layers up to 300 nm thick by using a high repetition rate laser that provided 5 ns pulses with fluences up to 0.9 J/cm^2.

It is always difficult to define the actual growth rate of the oxide films produced by laser irradiation, since the temperature varies considerably during the complete operation and it becomes difficult to define an appropriate timescale for the growth period during such a temperature cycle. Rough estimates [5.22] have produced oxidation rates many orders of magnitude faster than those obtainable in conventional solid-phase thermal oxidation of silicon (Chap.8). Similarly fast reaction rates can be calculated from the data of FOGARASSY et al. [5.31]. This group have also shown that oxygen incorporation by mul-

tishot melting in the UV is very much enhanced by the presence
of specific impurities during resolidification. In fact, in contrast to As and Sb, both In and Bi encourage significantly
higher rates of oxygen incorporation. In this case the mechanism is thought to be related to surface segregation effects
associated with each type of impurity. Further details on
impurity effects are discussed in Sect.5.4

Experiments by BENTINI et al. using a ruby laser [5.21,32]
have produced oxide films around 50 nm thick, with no difference in the amount of oxygen incorporated when various surface
preparations were used. The stoichiometry of the films grown
was close to that of silicon dioxide, with a ratio [O]/[Si] of
1.9. More refined estimates of the formation times involved
have resulted in proposed growth rates of 10^8 nm/s, which is
10^9 times faster than solid phase oxidation.

Clearly the fact that the silicon is well above the melting
point is significant for all of these reactions. This is quite
evident from Fig.5.3 which shows the amount of oxygen reacted

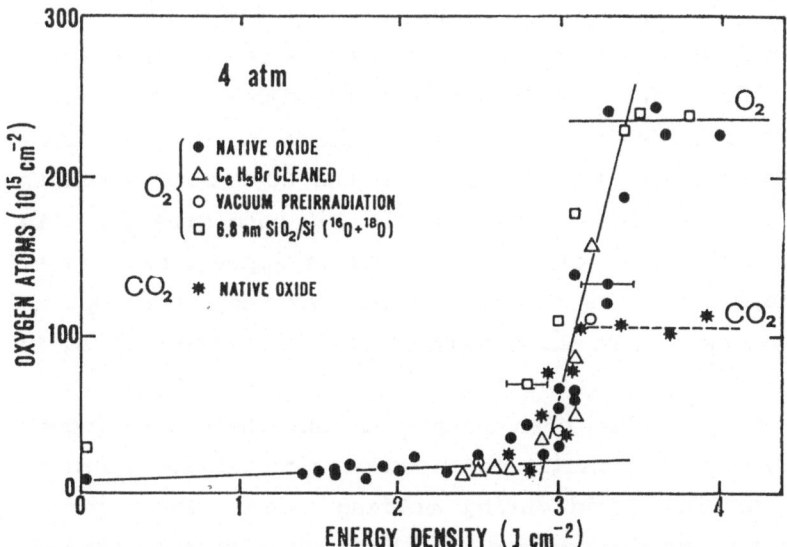

Fig.5.3. Plot of the amount of oxygen incorporated into the
silicon surface as a function of incident laser beam fluence
[5.32]

on the silicon surface as a function of energy density of the incident laser pulse, which in this case, was of 20 ns duration [5.32]. However, as discussed above, melting may not be the sole criterion for induced growth of thick oxide layers. For example, the data of Fig.5.3 indicate that appreciable oxygen uptake is not obtained until fluences of some 3 J/cm^2 are applied. This is known to be around 3 times the melting threshold of c-Si, and introduces the question as to what mechanism is induced by these very intense laser pulses.

The initial model for this mode of laser oxidation was the following [5.32]: The O_2 molecules impinge upon the liquid surface and dissociate, thereby supplying two oxygen atoms to form SiO_2. Since the diffusion coefficient of oxygen in liquid silicon is approximately 10^{-4} cm^2/s, it only takes about 100 ns for oxygen to diffuse 100 nm and the reaction will proceed very quickly until either (a) the intermediate oxide layer becomes so thick that it presents a strong barrier to diffusion of the oxygen species, and this becomes the limiting mechanism in the reaction, or (b) the number of silicon atoms available for the reaction near the silicon surface is significantly reduced.

The limiting availability of oxygen atoms for diffusion was checked by replacing the O_2 environment with CO_2. When each CO_2 molecule collides with the sample surface, only one atom will be supplied for diffusion, since the CO=O bond strength (127 kcal/mole) is almost equal to the O-O bond strength (119 kcal/mole), while that of the C-O bond is appreciably higher (257 kcal/mole). From Fig.5.3 it is seen that the oxygen uptake using CO_2 is indeed only about half of that measured when O_2 is present.

More recently, it has been suggested that the oxygen incorporation may be significantly enhanced once convective currents are set up in the liquid during melting [5.33]. The vigorous turbulent action at the surface will have the effect of replenishing the surface with fresh unsaturated silicon atoms eager to bond with the impinging oxygen. Without such a mechanism,

202

the liquid surface could rapidly become covered with several layers of an oxide "slush" which would otherwise prevent further penetration of oxygen, and oxide formation. Further complications to the process may also be introduced, at high beam intensities, by the ejection of a variety of species from the highly excited surface, both during and after irradiation, and their subsequent interactions with the incident oxygen. Clearly this is a highly nonequilibrium regime involving many nonlinear mechanisms, and as such will most likely not strongly compete against conventional processing techniques, or even the more controllable laser-assisted film-forming procedures.

5.3 Oxygen Implantation

Rather than depending on the availability of gaseous oxygen and its diffusion through already existing oxide layers to the appropriate reacting interface, it is also possible to actually implant oxygen into the silicon lattice and then anneal the impurity ions into substitutional lattice sites. In fact heavy implantation doses are being considered as a means by which buried oxide layers could be fabricated as a viable alternative to the rather expensive silicon on sapphire. Naturally, laser annealing has been considered in these schemes. However, a lighter implant of oxygen could be used to assist in a novel alternative low-temperature method of formation of silicon oxide i.e. directed ion-beam oxidation.

Oxide layers have been produced on c-Si by laser annealing samples that had been implanted with oxygen ions to a depth of $0.12 \ \mu m$ and a peak concentration of $8 \cdot 10^{21} \ cm^{-3}$ [5.34]. A poly-crystalline layer always regrew at the top of the undamaged crystal substrate after annealing, and above this an amorphous oxide film was formed, as seen by transmission electron micros-copy (TEM) [5.35]. Differential infrared spectrometry revealed the presence of Si-O bonding reminiscent of an SiO_2 structure.

SIMS depth profiling revealed that surface segregation of the oxygen occurred as a strong function of incident laser fluence. This was attributed to the usual zone-refining process which often occurs during rapid laser annealing, and which is therefore affected by the regrowth velocity. The formation of the polycrystalline region during annealing is most likely a result of the oxygen impurities inhibiting good epitaxial regrowth, as discussed earlier for the case of CW laser annealing of a-Si in O_2. Unfortunately, no data on the electrical properties are presently available for the films grown by this technique.

5.4 Nitrogen Implantation

Just as large dose oxygen implantation is being investigated with the aim of producing buried dielectric layers in single crystal silicon wafers, the prospects for nitrogen implantation in order to produce buried silicon nitride layers are also being studied [5.36].

Recent electron spin resonance measurements (ERS) [5.37-39] indicate that substitutional nitrogen impurities can not be incorporated into lattice sites either by laser annealing of N-implanted silicon, or by doping with nitrogen during the crystal growth. Infrared (IR) absorption studies have shown that local mode frequencies similar to those for silicon doped with nitrogen in the melt can be observed after implantation with nitrogen, without any annealing [5.40]. After conventional annealing at temperatures up to 600°C, the intensities of these absorption features actually increase. However, following annealing at 700°C, additional bands appear that are at positions similar to those seen for Si-N absorption in crystalline Si_3N_4, and it is suggested that at this temperature, precipitation of the implanted nitrogen actually occurs. Annealing at 900°C was found to remove all the bands associated with Si-N absorption.

Pulsed laser annealing of silicon implanted with N_2 or N^+ ions has been found to result in substitutional nitrogen impurities with neutral charge, $N^{(0)}$ being introduced into the silicon crystal [5.39]. These studies suggested that nitrogen could be introduced into substitutional lattice sites under conditions involving rapid cooling and recrystallization rates. CHIANG and co-workers [5.34] have performed similar experiments, using instead of the ruby laser employed by the abovementioned group, a Q-switched Nd:Glass laser. Again IR spectrometry revealed that Si_3N_4 could be formed, and this was confirmed by electron diffraction measurements.

In contrast to the case of oxygen implantation followed by pulsed laser annealing, with nitrogen cell-like structures always appeared on the surface of the treated layer. These structures are most likely due to the lateral rejection of the impurity, resulting from instabilities in the resolidification interface due to the occurrence of constitutional supercooling. This phenomenon usually occurs if the impurity element exceeds its usually low equilibrium solubility concentration in the silicon lattice. This effect of moderately heavy implants of N^+ will probably restrict the use of this method of forming silicon nitride material.

5.5 Impurity Effects

The effects of impurities during oxidation has been briefly touched upon in Sect.5.2. Recently, FOGARASSY et al. [5.36] have shown, as is the case for thermally grown silicon, that heavily doped substrates ($5 \cdot 10^{21}$ As/cm^3) oxidize significantly faster than pure samples when induced to oxidize in the solid phase with ArF-laser radiation. More excitingly, when irradiated in O_2 at fluences close to those expected to induce melting, an exceedingly strong dopant influence was noted. After about 10^4 repetitive ArF-laser pulses, samples previously implanted with

As, N_2 or F, readily oxidized to thicknesses up to 90 nm. On the other hand, for samples previously implanted with Si^+ ions, no incorporation of oxygen could be detected. Detailed studies of the redistribution of the impurities during and after irradiation as well as the specific chemical role of each species are clearly required. Preliminary RBS experiments in this area [5.36] reveal As redistribution both into the silicon and the oxide, whilst SIMS and Auger spectroscopy show that fluorine and nitrogen appear to diffuse through the oxide into the gaseous phase or accumulate at the $Si-SO_2$ interface.

Impurity effects have been highlighted by WHITE et al. [5.37] in conventional furnace oxidation on Ge-implanted Si. The group found that Ge was rejected from the growing SiO_2, and accumulated at the SiO_2-Si interface, occasionally growing epitaxially on the substrate. Furthermore, an enhanced oxidation rate was observed which was thought to be due to an increase in the interface kinetics of the reaction. Since the Si–Ge bond strength is smaller than that for Si-Si, it may be that the presence of Ge provides an easier route to making more Si bonds available for oxidation.

5.6 Laser Cleaning of Surfaces

Many areas of technology have today advanced so far that the presence of contaminants either during fabrication or application can contribute greatly toward the degradation of desirable properties and eventual device failure. The study of the physical and chemical properties of contaminated surfaces, however, relies fundamentally upon the ability to produce atomically clean surfaces to start with. It has been shown in recent years that pulsed lasers can be used to prepare such atomically clean surfaces in ultrahigh vacuum for the case of silicon [5.38,43,44].

Figure 5.4, for example, shows an Auger electron spectrum of an air-exposed silicon sample after baking in a UHV system.

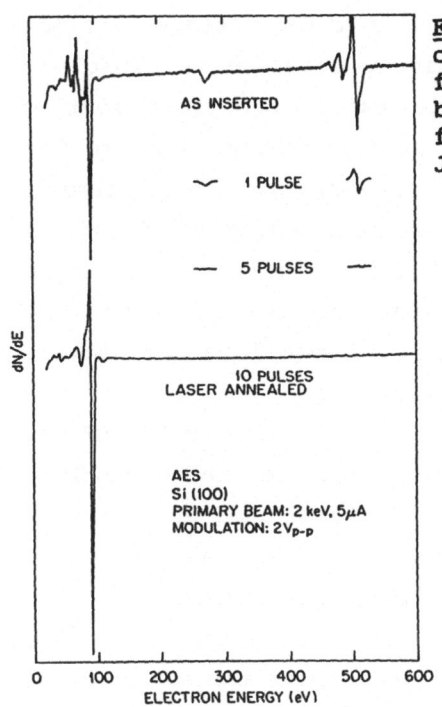

Fig.5.4. Auger electron spectrum of an uncleaned silicon <100> surface (top), and after irradiation by 1, 5, and 10 laser pulses at a fluence of approximately 2.0 J/cm² [5.43]

After five laser pulses at a fluence of 2 J/cm² (see bottom of Fig.5.4) the absence of the Auger peaks near 265 and 500 eV indicates the loss of carbon and oxygen impurities. Indeed the results indicate a reduction in the concentration of these atoms by a factor of 50 and 500, corresponding to surface concentrations of the order of 0.1 of a monolayer, or less. Additional photoelectron studies (PES) also indicate that the surface is essentially free of hydrogen impurities [5.29]. These reduced impurity levels which, incidentally, are obtained in a matter of a small fraction of a second, are comparable to those achieved by repeated sputtering and conventional thermal annealing over a time period more usually on a scale of days.

This surface cleaning effect was found to have a threshold under the present conditions of about 1 J/cm². It was also found that as the number of pulses increased, the more complete was the removal of the contamination. There are two possible explanations for the removal of the chemical impurities.

Firstly, they may simply be desorbed from the surface duration irradiation into the UHV. Alternatively they may be dissolved into the bulk of the material as a result of surface melting (the threshold value for cleaning is remarkably close to the single shot melting threshold of silicon for these pulses). Furthermore, complete and uniform redistribution of the atoms over a depth equivalent to the melt-front penetration would be expected to give rise to a surface concentration of about 0.3% of a monolayer of oxygen and 0.1% of a monolayer of carbon, remarkably close to the observed quantities.

WESTENDORP et al. [5.30], using resonant nuclear techniques in conjunction with RBS, and AES, concluded that there was less than 1% interdiffusion of oxygen during laser pulses that resulted in cleaning of the surface, and also that the concentration in the bulk was less than the solubility limit in the case of oxygen. The absence of indiffusion was explained by noting that it would take almost 1 second to dissolve the native oxide layer of the silicon (approximately 2 nm), at the known dissolution rate of 10^{-5} cm/min, into the melted zone. To add weight to this interpretation, it is also known that a strong pressure increase is always observed in the irradiation chamber immediately after irradiation. These observations strongly indicate therefore that desorption is a major factor in the surface cleaning effect. Although no wavelength dependence has been performed, it is expected that the desorption mechanism is purely thermal in nature, with no bond selectively in the process.

6. Laser-Induced Deposition

In this chapter film growth from the gas phase is described. Two main reaction modes are possible, namely, laser pyrolysis and laser photolysis. The former is closely related to traditional chemical vapour deposition, but with use of lasers significantly enhanced reaction rates can be achieved. Laser photolysis utilises selective photochemical bond-breaking, so that much reduced processing temperatures can be employed. Oxide and nitride deposition from various precursor mixtures, as well as organic polymer formation, are reviewed.

6.1 Background

In conventional chemical vapour deposition (CVD) the constituents of the vapour phase, which are quite often diluted with an inert carrier gas, react chemically at a hot surface to form a deposited solid film, while the by-products are carried away in the gas phase. The attraction of CVD lies not only in its versatility for depositing a great variety of compound and elemental films on many types of substrate, but also in the fact that these substrates can be held at relatively low temperatures, typically 600°C or less. Additionally, the structure of these films can often be altered by varying the deposition conditions, such as temperature, gas mix, gas pressure etc., and whether it be epitaxial growth, amorphous or polycrystalline films that are required, the layers can usually be deposited with a large degree of purity and control.

The most common CVD techniques are atmospheric CVD, low pressure CVD (LPCVD), and plasma assisted CVD (PCVD), sometimes referred to as plasma deposition [6.1,2]. Although most effort has centred on low-temperature processes, deposition conditions have been known to vary from as low as 100°C, to as high as 1000°C, while gas pressures can range from 1 atmosphere (760 Torr) down to 7 Pa (0.05 Torr). Four main types of reactor can be employed [6.3]. Basically these are:

(a) Hot-wall, reduced pressure reactor,

(b) continuous atmospheric pressure reactor,

(c) parallel-plate plasma deposition reactor, and

(d) hot-wall plasma deposition reactor.

The atmospheric pressure reactors tend to have low throughputs, require extensive sample handling, and provide film uniformities that are no better than ± 10%. On the other hand, LPCVD offers very fast deposition rates, high throughputs, and much more uniform step coverage. In general, though, the film properties tend to improve with increasing substrate temperature, as will be discussed later in this chapter. Plasma assisted techniques [6.4] provide deposited layers at lower substrate temperatures in general. However, the film properties tend to be slightly poorer than those prepared by the best atmospheric or low temperature methods. Table 6.1 summarizes the current state of affairs in this area of film deposition [6.3].

Many other methods for dielectric film deposition have been and are currently being explored to various degrees. These include anodization, evaporation, molecular beam epitaxy, photochemical deposition, and various types of sputtering. Here we concentrate on the various laser-based techniques for depositing thin layers. This class of reaction is known as laser CVD, or LCVD, and can operate either via photochemically induced reactions, or by pyrolytically controlled processes most analogous to conventional CVD. As described in more detail in Chap.2, photochemical deposition will occur when the incident

Table 6.1. Comparison of different depostion methods for insulating materials [6.3]

Deposition properties	Methods			
	Atmospheric pressure CVD	Low-temperature LPCVD	Medium temperature LPCVD	Plasma-assisted CVD
Temperature(^{o}C)	300–500	300–500	500–900	100–350
Materials	SiO_2 P-glass	SiO_2 P-glass	Si_3N_4 SiO_2 P-glass	SiN SiO_2
Uses	Passivation, insulation	Passivation, insulation	Passivation, insulation	Passivation, insulation
Throughput	High	High	High	Low
Step coverage	Poor	Poor	Conformal	Poor
Particles	Many	Few	Few	Many
Film properties	Good	Good	Excellent	Poor
Low temperature	Yes	Yes	No	Yes

photons from the laser beam are resonantly absorbed by one or more of the gas phase species to such a degree that the molecular bonds are ruptured creating a reactive species that initiates the chemical reaction near a solid surface. This process is clearly dependent upon the absorption cross-section of the donor-bearing molecules, the gas mixture and pressure, as well as the intensity and wavelength of the laser radiation. Because of the important interplay between wavelength and molecule, the reaction can be considered to be selective in nature. Since it is the gas phase that is excited by the radiation, the substrate need not be heated up to particularly high temperatures, and this is probably one of the most desirable characteristics of this class of laser processing.

Pyrolylis is quite similar to the usual CVD processes. In this case the substrate is rapidly heated up to the desired temperature over a particular area, and the reactant atoms are liberated from their parent molecules when they decompose by collisional excitation with the hot solid surface. Since the mechanism must be initiated by strong absorption of the laser radiation in the substrate, it is necessary to match the wavelength of the radiation to the optical properties of the solid. As has been pointed out in Chap.2, there is a much wider opportunity for absorbing radiation in solids than there is for isolated molecules because of the number of energy states available to the electrons. Indeed the band structure exhibited by most solids usually presents the user with a wide choice of laser wavelengths for heating up the material to the desired temperature.

The role of surface adsorption can also be important in dissociating the precursor molecules. There are quite often conditions where the gas-phase species stick to the substrate and become physi- or chemisorbed. In these states they can also be photochemically or pyrolytically dissociated either directly by absorption in the surface-adsorbent system or indirectly through heating of the substrate or adsorbate.

The deposition of thin films is also possible by means of laser evaporation [6.5]. In this case, an intense source of laser radiation is employed to evaporate from a target, material which is then deposited on to a substrate positioned some distance away. This technique has recently been applied to a wide variety of solid target materials, and seems to be an extremely versatile method for producing a very large variety of thin film coatings. Multiple targets can be simultaneously irradiated, and in this way various new mixtures of films can be prepared. This is potentially a very clean and fast process, and again since strong absorption and heating (plus evaporation) are necessary, a very wide range of source materials in the solid phase is available. Recent experiments at AT&T Bell

Laboratories have used the laser evaporation technique to produce so-called "high-temperature" superconducting material.

This chapter is divided into sections dealing principally with the deposition of metal oxides, with semiconductor oxides and dielectrics, and with organic films. Tables summarizing the various materials deposited by these methods are shown in Chap.4.

6.2 Metal Oxides

One of the earlier publications in this area described the deposition of TiO_2 using $TiCl_4$ and H_2O which had been obtained by reacting CO_2 and H_2 [6.6]. In order that the substrate can be adequately heated to stimulate the required reaction, the donor molecules must be sufficiently transparent to the incident radiation, and the substrate, in this case quartz, must be an efficient absorber of the radiation. Since $TiCl_4$ is transparent at the CO_2 laser deposition wavelength, and quartz ($c-SiO_2$) is a strong absorber of 10 μm radiation both criteria are easily met. It was found that the oxide thickness was approximately a linear function of irradiation time, with no self-limiting effects being obvious. Deposition rates up to 20 μm/min were observed although the actual composition of the films varied somewhat in nature over the diameter of the larger deposited spots. The films were in general optically clear and could not be removed by vigorous scrubbing. This is to be compared with TiO_2 films deposited by conventional CVD which are often full of pinholes and have very poor adherence.

It is also possible, by carefully choosing the donor molecules and the appropriate wavelength, to encourage the gas mixture to resonantly absorb the photonic energy and subsequently dissociate. Since chemical bonds must therefore be broken, appreciable absorption of energy must occur, and, not surprisingly, the use of powerful ultraviolet lasers is predominant in this field. Energetic photons are, however, not always a necessary prerequisite. It is sometimes possible to "crack" large donor-bearing molecules using multiple absorption of much lower energy infrared photons. Thus, multiphoton dissociation is sometimes a possibility, and since "multi" is an essential characteristic of the process, very large photon fluxes are generally required.

Direct photon stimulation of various gas phase or surface adsorbed molecules by UV laser photons has successfully resulted in the deposition of thin films via reactions such as

$$Zn(CH_3)_2 + NO_2 \text{ or } N_2O + h\nu \rightarrow ZnO + \text{products} , \tag{6.1}$$
$$Al_2(CH_3)_6 + NO_2 + h\nu \rightarrow Al_2O_3 + \text{products} , \tag{6.2}$$
$$(CH_3)_3 \text{ InP } (CH_3)_3 + O_2 \text{ or } H_2O + nh\nu \rightarrow In_2O_3 + \text{products} , \tag{6.3}$$
$$CrO_2Cl_2 + h\nu \rightarrow CrO_2Cl + Cl \rightarrow CrO_2 + Cl + Cl . \tag{6.4}$$

During the deposition of InP films on various substrates using excimer laser photodissociation of $(CH_3)_3$ InP $(CH_3)_3$ and $P(CH_3)_3$ it has been found that when trace amounts of oxygen are present indium oxide layers can be formed [6.7]. Both ellipsometry and Auger Electron Spectroscopy (AES) have shown that stoichiometries close to In_2O_3 can be produced. Under these conditions the beam was incident approximately parallel to the substrate (InP) at a glancing angle of approximately 5 degrees. The actual refractive index of the deposited material was 1.89 compared to 2.0 for true In_2O_3. A deposition rate of approximately 0.02 nm/laser pulse, each of 15 ns duration, was obtained, and film thicknesses up to 134 nm were grown. No appreciable carbon or phosphorous contamination could be detected. When irradiation was performed perpendicular to the substrate, a film with a metallic appearance was obtained. This turned out to be InO with an almost exact 1:1 ratio of In and O. Again there were no measurable C or P impurities.

Much attention has been paid to aluminium oxide films in recent years because of their ability to efficiently restrict the diffusion of particularly mobile impurities, especially sodium ions. Al_2O_3 also exhibits an especially good resistance to radiation damage. In fact these films are superior to SiO_2 in both of these characteristics. Also, like Si_3N_4 it has a larger dielectric constant than SiO_2. Several groups have studied the laser photolytic deposition of Al_2O_3 using high power excimer lasers [6.8,9]. Using the 248 nm line of the KrF laser, SOLANKI et al. [6.8] have deposited uniform films of Al_2O_3 at rates of approximately 200 nm/min over complete 7.5 cm wafers, using 80 mTorr of trimethylaluminium (TMA) and a total cell pressure of 1 Torr. The He flow rate was 100 sccm[1] and the N_2O flow rate varied between 50 and 1000 sccm. Under the same conditions increased deposition rates were obtained when the 193 nm line (ArF) was used, but the films were not so uniform. Only by reducing the TMA partial pressure to 30 mTorr could acceptable uniformity be restored. Growth rates of over 1 μm/min were achievable, but the films peeled off the substrate when exposed to the atmosphere, and some became powdery in appearance at the highest deposition rates. The best films were grown with a N_2O flow rate of 200–800 sccm and a TMA partial pressure of 80–120 mTorr. By changing the ratio of the input reactants as well as the wavelength of the incident radiation the stoichiometry of the films could be altered. The electrical and physical properties of the films were very similar to layers produced by magnetron RF sputtering with notable exceptions that the density of pinhole defects was significantly reduced (by more than a factor of 30) in the laser CVD material and that the electrical resistivity was also reduced by an order of magnitude.

The detailed mechanism of the photochemical reaction in forming these layers is not yet well understood. For example, although in these experiments the films were principally depo-

---

[1] Note that the units sccm is commonly used to quantify gas flow; sccm stands for standard cubic centimeter per minute, thus defining a flux of one cm^3 of gas per minute at 273 K and 760 Torr.

sited using a beam passing above and parallel to the substrate, there is a well documented report of modifications to the reaction when photons impinge directly onto the substrate. For example, DEUTSCH et al. [6.10] stated that the additional energy incident on the substrate would not be sufficient to cause transient heating that would be significant compared to the substrate temperatures used. (It was estimated that incident fluences around 1 mJ/cm^2 will raise the temperature of the silicon surface by less than 10°C). A number of possible additional photochemical mechanisms have been suggested to account for the enhanced effects observed. These include dissociation of clusters that are formed in the regions above the substrate which are normally outside the area directly in the UV beam line, nucleation of the surface by the photolysis of the adlayers, or photoenhancement of the mobility of particular surface species. It was particularly postulated that so long as the UV radiation can couple into the surface-adatom system, the amount of photon energy available (6.4 eV) should be sufficient to appreciably increase the surface mobility of, for example, Al adlayers [6.11]. The technique of applying direct UV radiation to the substrate during deposition is also reported to improve the physical and electrical properties of Al_2O_3 films.

Aluminium oxynitride (AlON) has also been laser deposited on substrates held at 300°C using an ArF laser, for potential application to InP MIS devices [6.12]. (Traditional depositional methods are thought to induce surface damage resulting in current drift and interfacial problems). Films up to 80 nm thick have been grown during irradiation periods of around 18 min, and it was found that an inversion layer could be formed in devices incorporating such a dielectric. Very good quality films of phosphorous nitride (PN) on InP substrates have been produced by HIROTA and MIKAMI [6.13] using pulsed excimer laser CVD. The deposition temperature was reduced from the levels around 500°C required by conventional thermal CVD to around 300°C in the laser CVD technique whilst the same mixtures of

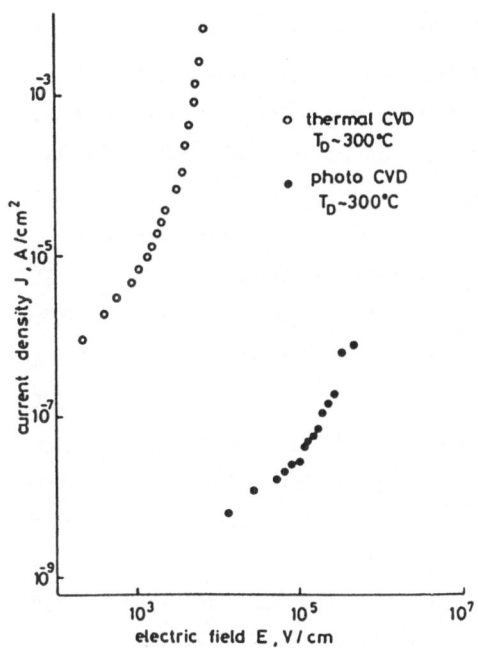

Fig.6.1. Plot of the current density against electric field for n-type InP film structures of thickness 120 nm and area $1.1 \cdot 10^{-3}$ cm^2, fabricated by thermal CVD and photo CVD [6.13]

o thermal CVD $T_D \sim 300°C$

• photo CVD $T_D \sim 300°C$

current density J, A/cm^2

electric field E, V/cm

phosphine (PH$_3$) and ammonia (NH$_3$) were used. Films formed at lower temperatures, around 200°C exhibited gradual hygroscopic deterioration in air after preparation. Deposition rates of typically 40 nm/min were achieved under the most ideal conditions and film thicknesses of approximately 120 nm were grown. These photodeposited films exhibited a much larger resistivity and lower leakage current than did the thermally prepared layers, as shown by the plot of current density against applied electric field in Fig.6.1. It is quite evident that this is a most promising method of preparing this little-used insulating material, which is expected to become an excellent gate insulator for InP MIS devices.

There is considerable interest in producing uniform large area films of the wide bandgap semiconductor ZnO due to its transparency down to below 400 nm and well into the infrared, its photoconductivity, birefringence and other desirable properties. Being less expensive than indium tin oxide (ITO) it has attracted interest for heterojunction solar cells. LCVD allows rapid ($\simeq$300 nm/min) deposition of ZnO over 2 cm by 5 cm areas [6.14] using dimethylzinc (DMZ) mixed with NO_2. When N_2O rather than NO_2 provided the oxygen for the reaction, there was an appreciable reduction in the deposition rate ($\simeq$20 nm/min), and this was generally unchanged when 193 nm photons from the ArF laser rather than 248 nm photons from the KrF laser were present, despite the absorption peak of DMZ being near 250 nm. It is reported that the best results with respect to deposition rate and stoichiometry were obtained with 30 mTorr of DMZ, a NO_2 flow rate of 34 sccm, and a buffer He flow rate of 100 sccm, with an overall cell pressure of 2 Torr [6.14]. The deposited films appeared clear, with no measurable trace of N and less than 1% carbon contamination, and were uniform from end to end to within 5%.

Highly stoichiometric films (49% Zn, 51% O) were obtained under optimum conditions. However, by increasing the ratio of DMZ/NO_2 flow, the stoichiometry could be easily altered, causing the resistivity of 0.5 μm thick films to vary from 10^{-1} to 10^3 Ωcm. The refractive index of the stoichiometric film deposited at 200ºC was 1.86 at a wavelength of 632.8 nm, although this was found to increase with higher deposition temperatures and increasing abundance of Zn in the film. During adhesion tests the silicon substrate always cracked before the ZnO could be detached, while the pinhole density of a typical 100 nm thick layer was measured to be <1 per cm^2. Highly oriented ZnO films have also been grown using lasers in a completely different mode of operation than that described so far. We have seen that with laser controlled heating one must find a laser wavelength to heat up the actual substrate upon which the films will be grown. There are limitations on the types of gases which may be present during this process in that they must not significantly absorb the incident radiation. There may also be disadvantages in heating up substrates for prolonged periods of time. Alternatively, laser photolytic CVD can relax the choice of substrate somewhat since significant heating is not required. However, there are presently constraints on the process through limitations on the availability of the appropriate gas phase molecules with the necessary reactant bearing constituents.

Mixed films of Cr_2O_3/CrO_2 up to several microns in thickness, as well as 1 mm long single crystals of Cr_2O_3 exceeding heights of 1 mm have been photolytically grown at rates up to 3 μm/s, using an argon laser operating in the 488–514 nm region [6.15]. This appears to be the first report of laser photochemically induced growth of single crystals of a binary compound. The reaction is thought to be initiated as given in (6.4), with the adlayers contributing significantly to the reaction for low beam powers. Once the reaction has been initiated photochemically by higher beam powers, pyrolytic effects are considered important. In the presence of a reducing reaction, the same group reported an increase in the Cr:O ratio from 2:3 with 0.1 Torr $Cr_2O_2Cl_2$ to 6:5 in the presence of 200 Torr HCl, using an ArF laser [6.15].

The novel technique of laser evaporation offers some of the advantages of both of the above processes [6.16]. Laser evaporation is basically a high vacuum technique whereby the energy for evaporation is delivered from an external laser source focused onto a pellet or pellets of the appropriate atomic composition (Fig.3.7) The energy supplied from intense CO_2 laser pulses is sufficient to evaporate compounds in such a short time that thermal diffusion within the target is not appreciable and fractionation of the individual constituents is relatively negligible. This process is essentially the same as conventional "flash" evaporation which is more usually initiated using other energy sources. The substrates on which the evaporated atoms condense are placed nearby and of course there is little restriction on their physical nature. Therefore a wide variety of substrates, Si, GaAs, quartz, Corning 7059 glass, gold and titanium films, and sapphire have been used to study the deposited layers.

The crystallinity of ZnO layers prepared by laser evaporation was found to be dependent on the substrate type and temperature, and generally improved as the temperature was increased and the growth rate reduced. Oriented polycrystalline layers were even grown at temperatures as low as 50°–100°C, which is lower than any previously reported temperature for the growth of ZnO films. The refractive index of the layers was measured to be in the range 1.98–2.04, compared with the value of 1.86 obtained by the laser CVD technique described earlier. The films were n–type and conducting (resistivity of 0.01–1.0 Ωcm, but by annealing in an O_2–rich environment for 60 min they could be made highly resistive (resistivity of 10^6 Ωcm).

Experimentally, the growth rate was found to depend inversely on the substrate temperature for a given evaporation rate. The growth of the films was thought to occur by reactive deposition of Zn and O_2 on the substrate and a variation of the sticking coefficient of Zn at the ZnO surface was the reason for the reduction in growth rate with temperature. Typical deposition rates of 0.1 nm/s were measured, but growth rates of up to 2 nm/s were attained with a slight loss of stoichiometry which was attributed to rapid oxidation of Zn adatoms. These growth rates are almost an order of magnitude, at worst, and only a factor of 3 or so at best, slower than those previously obtained by laser CVD. However, it appears that this technique may offer more in the way of flexibility than LCVD as well as films of slightly better quality making it a more attractive proposition. In fact it has been stated by SANKUR and CHEUNG [6.5] that further optimization of the technique is expected to yield substantially improved ZnO films for device applications.

The unique advantages of laser evaporation have not been limited to the production of ZnO films. It has also been applied to the formation of many oxide, fluoride and semiconducting layers [6.9]. This is principally because, as mentioned earlier, pulsed lasers allow congruent evaporation of compounds or mixtures which in most cases allow the source stoichiometry to be faithfully reproduced in the deposited films.

It is worth pointing out at this stage that CW lasers have also been used in the area of laser assisted evaporation. Just as pulsed laser applications in this field were equated to the process of flash evaporation, the CW laser technique has been described to be most akin to conventional thermal evaporation [6.5]. The evaporation products of ZnO have in fact been analysed for both methods using a residual gas analyser. Whereas neutral atoms of Zn and O_2 were observed when the CW laser was used, a substantial amount of Zn ions (10^3 more than Zn atoms) ozone (O_3), and both atomic and excited O_2 were detected, indicative of probable extra-thermal mechanisms. This is obviously an area requiring much more study, but nevertheless one with a great deal of potential.

Irradiation of metals by intense laser pulses can also lead to the generation of a localized plasma just above the surface, which can initiate excited reactants that may combine with the metal atoms to form overlayers as thick as 1-5 nm. This recently applied surface chemistry technique has been called Laser Pulsed Plasma Chemistry (LLPC) by MARKS et al. [6.17,18], and has in fact been used to form oxide layers on niobium surfaces under highly controlled conditions. This is to be strictly separated from the usual laser-induced heating of the surface where thermal mechanisms are considered to be dominant [6.19]. The plasma conditions necessary to invoke this mechanism can be relatively easily initiated by pulses from a CO_2 laser. Irradiation of the metallic surface by pulses from an excimer laser, however, does not produce the required conditions, possibly because of the much higher plasma initiation threshold at the shorter wavelengths, as well as the optical absorption properties of the surfaces layers. Furthermore, once any plasma is formed it will absorb the infrared radiation extremely efficiently, and hence if the CO_2 laser pulse has a typical "tail" characteristic then this will be coupled strongly into the

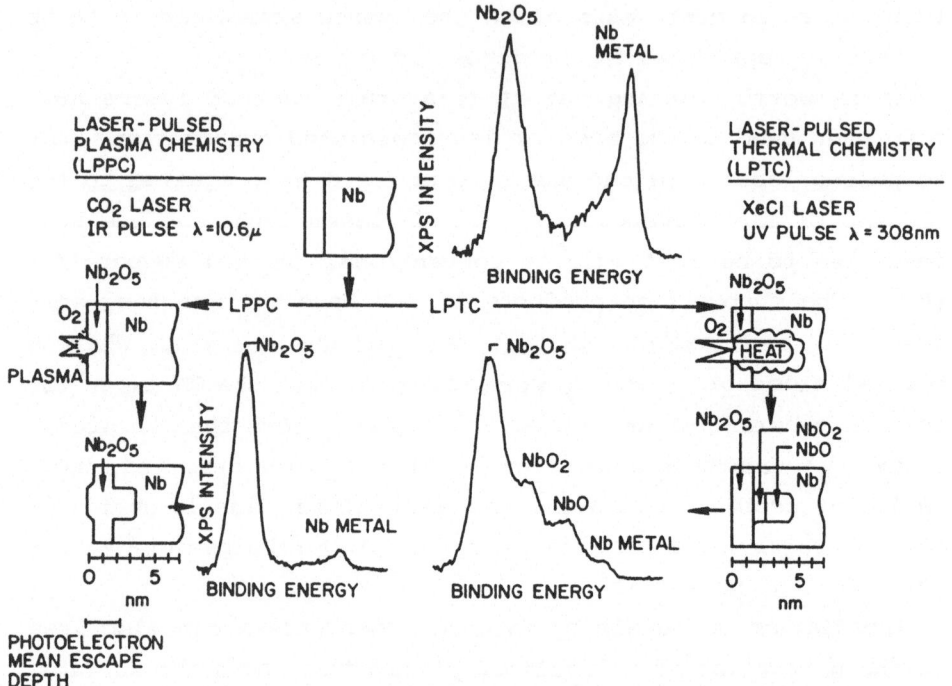

Fig.6.2. Schematic representation of the oxidation of niobium by LPPC and LPCT techniques. A well defined stoichiometry and a sharp oxide/metal interface is formed by the former method, while the latter thermally controlled process produces a graded stoichiometry, as shown by the accompanying XPS spectra [6.18]

plasma, causing further excitation. XeCl laser pulses then, will induce oxidation mainly by heating the lattice as mentioned previously, and the oxides will resemble those grown by the usual thermal techniques.

These two processes are shown in greater detail in Fig.6.2. As can be seen, with LPPC only Nb_2O_5 is formed, with a generally much smaller valence defect δ than thermal or RF plasma grown layers.[2] Thicker films could always be grown by using only one pulse rather than multiple pulses. In fact, the oxide layers usually got progressively thinner with increasing number

---

[2] The valence defect δ in $Nb_2O_{5-\delta}$ is considered to be a measure of incomplete oxidation.

of pulses. This was probably due to the obstruction of plasma formation once the first pulse has formed a thin oxide layer and reduced the electron tunnelling cross-section from the metal to the gas/solid interface. Simultaneously, there will be increased absorption of radiation by the newly grown oxide layer, which will tend to increase the rate of evaporation of the film. This effect gives rise to a self-limiting process of oxide growth, a rare and potentially useful feature in thin film preparation.

6.3 Silicon Oxide

Although thermal oxidation of silicon in carefully maintained RF heated quartz tube furnaces usually produces the highest quality oxide layers available, there are applications where SiO_2 layers must either be produced at reduced temperatures or where highest quality is not essential to the application. In these cases the silicon dioxide layers are usually deposited onto relatively cool substrates using pyrolytically-controlled CVD reactions.

Traditionally, silica films are deposited by the pyrolytic oxidation of various alkoxysilanes such as tetraethylorthosilane (TEOS), or silane (SiH_4) via the following reactions:

$$Si(C_2H_5O)_4 + 12O_2 \rightarrow SiO_2 + 8CO_2 + 10H_2O \qquad (6.5)$$

in a cold wall CVD system around 800°C, and

$$SiH_4 + 2O_2 \rightarrow SiO_2 + 2H_2O \quad \text{at } 600°{-}1000°C , \qquad (6.6)$$
$$SiH_4 + O_2 \rightarrow SiO_2 + 2H_2 \quad \text{at } 300°{-}500°C . \qquad (6.7)$$

However, secondary reactions often occur during the oxidation of alkoxysilanes, and the presence of carbon, silicon monoxide, and various organic radicals, as well as water, significantly degrades the quality of the oxide layers. SiO_2 formation from silane is preferred at reduced temperatures for similar reasons, and has become the most commonly followed route to obtaining high quality deposited films. The reaction proceeds by the strong adsorption of oxygen on to the silicon surface where it subsequently reacts with silane to form silicon dioxide. Typical growth rates around 50-100 nm/min are achieved, while the $O_2:SiH_4$ mole ratio should not exceed 10:1 [6.20]. It is also usual to use diluted silane, commonly 5-10% by volume in argon or nitrogen.

More recently, photochemical decomposition using mercury photosensitization has developed as an alternative low temperature technology for preparing similarly large area silica films [6.21-23]. One of the first schemes identified proceeds as

$$Hg + h\nu \rightarrow Hg^* , \qquad (6.8a)$$

$$Hg^* + N_2O \rightarrow N_2 + O(3P) + Hg , \qquad (6.8b)$$

$$2O + SiH_4 + h\nu \rightarrow SiO_2 + 2H_2 , \qquad (6.8c)$$

and a low pressure mercury lamp as a source of incoherent UV radiation [6.21]. DIMITRIOU [6.22] has recently shown that photosensitization may even promote surface reactions during optical excitation of the silane/nitrous oxide system diluted with mercury.

Direct photodeposition of SiO_2 using conventional radiation sources but eliminating the requirement for sensitization has been reported for the SiH_4/N_2O mixture [6.24], and also for the systems:

$$SiH_4 + O_2 \rightarrow SiO_2 + products , \qquad (6.9a)$$

$$Si_2H_6 + O_2 \rightarrow SiO_2 + products . \qquad (6.9b)$$

In the former, when low pressures of O_2 were used, it was believed that photodesorption of the previously adsorbed oxygen molecules was the principal cause of the observed modifications to the thermal CVD reaction [6.25]. At higher partial pressures of oxygen no thermal CVD occurred and direct photo-lysis of the O_2 in the gas phase induced the reaction with the SiH_4, and SiO_2 was deposited. For (6.9b), enhanced growth rates (approaching 100 nm/min), and an observable stability increase over thermal CVD films grown using the same gaseous system, together with a complete absence of any Hg contamination [6.26], indicates that a promising low-temperature alternative to thermochemical deposition of silica may be at hand.

The capability of confining these photolytic reactions to a predetermined region in the reaction chamber using the larger energy densities available with lasers, enables much higher efficiencies and growth rates to be attained. A typical scheme used for photodeposition of thin layers, can use either a directly incident beam, or one which passes parallel to the substrate, as discussed earlier in Chap.3. BOYER et al.[6.27,28] first demonstrated the use of an ArF laser operating at 193 nm to photolytically deposit SiO_2 films as thick as 500 nm at dep-osition rates up to 5 nm/s but typically around 1-2 nm/s. Even with careful minimization of direct UV impingement on the sub-strate these growth rates are more than 20 times those achi-eved by using incoherent mercury lamp sources. Mixtures of N_2O and SiH_4 were used, sometimes with He or N_2 buffer gases added, and the deposition pressure was typically < 6 Torr. The LCVD SiO_2 deposition rate did not vary with substrate temperature up to nearly 600°C, primarily because substantial thermal reac-tions do not occur in N_2O and dilute silane mixtures under these conditions. The deposition rate did however increase as expected with increasing beam intensity. The mechanical, chemi-cal, and electrical properties of the films were found to be comparable to those produced by plasma and photosensitized layers, and in general increasing the substrate temperature improved the quality of the films in terms of pinhole density, hydrogen bonding, and reduced strain (or increased density). The electrical properties of the films will be discussed later in Chap.8. The use of perpendicular surface irradiation during the reaction has a large effect on the film properties, as was the case for the Al_2O_3 films [6.8] discussed earlier. In the present case, a large reduction in the amount of SiH and SiOH bonding in the films was observed under these conditions, and this had the effect of reducing the chemical etch rate and slightly increasing the refractive index. It was suggested that as well as stimulating the surface reactions, the direct irradiation of the surface provided in situ annealing of the layers during their growth. As Fig.6.3 shows, conformal step coverage of the films can be achieved over a 400 nm polysilicon step formed on an oxidized silicon wafer.

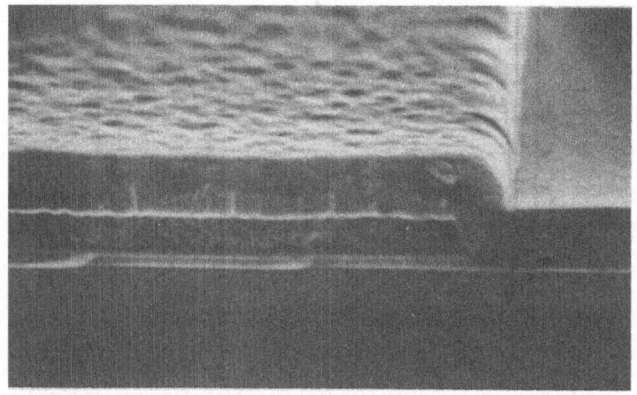

Fig.6.3. The conformal step coverage of silicon dioxide photochemically deposited over a 400 nm polycrystalline step [6.32]

The KrF laser (249 nm) has more recently been applied to photo-deposit SiO_2 layers, but from mixtures of SiH_4, O_2, and N_2 at typically 20 Torr and around 200°C [6.29]. Although SiH_4 only moderately absorbs at this wavelength, Si particles could be photo-deposited from pure silane using this laser at gas pressures around 100 Torr. Using a parallel beam configuration, oxide was deposited when all three gases were present in the reaction cell, and growth rates of 30 nm/min were obtained at substrate temperatures of 250°C. Between temperatures of 160 and 250°C, an activation energy of 0.18 eV was extracted from

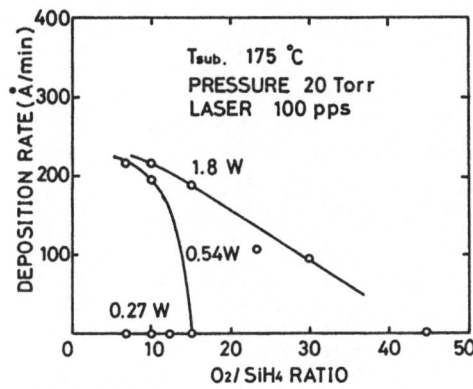

Fig.6.4 Deposition rate of SiO$_2$ as a function of the O$_2$/SiH$_4$ ratio for several laser beam powers [6.29]

the data [6.29], compared with a value of 0.22 eV for Hg lamp-induced deposition [6.24], and 0.13 eV for Hg sensitised CVD [6.23].

NISHINO et al. [6.29] have measured the dependence of the gas ratio of O$_2$/SiH$_4$ on the deposition rate for several beam powers for a substrate temperature of 175ºC (Fig.6.4). At 0.27 W, no deposition is achieved at any gas ratio used. At 0.54W the initially rapid deposition rate diminished significantly for higher ratios of oxygen to silane, while for 1.8W deposition stopped for ratios above 45. Furthermore, for any fixed ratio of gas, the oxide deposition rate tended to saturate. These observations suggest that the reaction may be limited by surface effects of the adsorbed species.

Further experimentation showed that at 175ºC and a gas ratio of 6.7, deposition did not occur without laser radiation. However, once some SiO$_2$ was formed by laser deposition (even for only 30 s at 1.8 W), the reaction could proceed at the same rate without the light. This study indicates that not only are photo-stimulated processes present at 249 nm, but that some hitherto unobserved laser-induced surface pre-nucleation processes are also operative.

Figure 6.5 shows a rod of SiO$_2$ grown by another LCVD process [6.30]. In this case the layer was deposited using a krypton laser beam to locally heat up the substrate in the presence of N$_2$O and SiH$_4$, the same reactive species used by BOYER et al. in

227

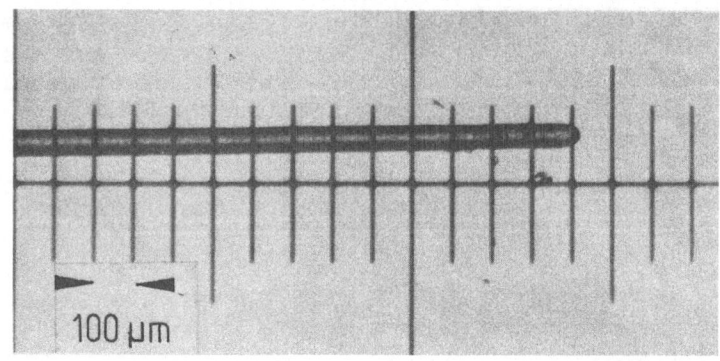

Fig.6.5. A rod of silicon dioxide produced by pyrolytic deposition [6.30]

the laser photodeposition technique. However in the present case, where partial pressures of 53 and 1-3 mbar, respectively, were used, no explosive gas phase reactions were observed for any of the irradiances used. The reaction is initiated by strong heating of the silicon upon which the gas phase donor molecules pyrolytically decompose and react. The oxide grown by this mechanism naturally remains at elevated temperatures and contributes itself to the absorption of the radiation, such that the hot tip of the rod (when the beam is slowly scanned) can activate the reaction. The films were found to be transparent, and typical for amorphous SiO_2 layers, in their Raman spectra and x-ray scattering, properties. Deposition rates of up to 2 $\mu m/s$, were achieved. Since the gases used here are essentially transparent to the radiation applied, relatively high gas pressures can be employed, unlike the case of laser photodeposition where there is the constraint that the gases must be optically transparent only to some small extent in order to preserve deposition uniformity.

Pyrolytic CVD of SiO_2 has more recently been achieved using either CW CO_2 or Ar ion lasers [6.31]. Since high surface temperatures were necessary, strong coupling of the incident photons to the substrate was required, and quartz and Si substrates were used respectively with each laser. Unlike the

lines deposited previously, the group were interested in preparing microlenses, and therefore spots were grown. For the
mixture of SiH_4 and O_2 used, the oxide deposition rates were
almost independent of temperature above a certain value, and
then at still higher values the rate decreased sharply. As a
consequence of this, the power in the Gaussian laser beam could
be arranged so that nearly flat disc-like shapes of deposit
were grown. By comparison, when SiH_4 and N_2O were used as
source gases, only mounds of oxide were obtained, indicative of
Arrhenius behaviour over a much wider temperature range. As
will be discussed in the following section, silicon nitride
films were also similarly seposited using this technique.

6.4 Silicon Nitride

Silicon nitride (Si_3N_4) has found many important applications as
an alternative to SiO_2 because it serves as an excellent barrier to the migration of the alkali ions, is nearly impervious
to water, and can also be used as an oxidation mask to make
planar structures. Because of its inherently high dielectric
constant (7.5 compared with 3.9 for SiO_2), nitride layers can be
proportionally thinner than their oxide counterparts. Si_3N_4
layers are commonly formed by chemical vapour deposition via
the reaction of ammonia (NH_3) with silane (SiH_4) or silicon tet-
rachloride ($SiCl_4$) or even silicon tetrabromide ($SiBr_4$) at temperatures between 700° and 900°C. Sputtering of silicon nitride
has not found widespread application in the microelectronics
area because of the difficulties encountered with reproducibility of important and necessary properties. In recent years
films of silicon nitride have been produced using plasma deposition techniques which offer the obvious advantage of being
strictly low temperature processes (200°–350°C). These films
are not strictly Si_3N_4, but are more commonly called plasma
nitrides (SiN). The properties of these layers however are not
yet of the standard required to see the technique introduced

into the production lines. Table 4.6 summarizes some of the preparation methods for these and other nitride films previously discussed in Chaps.5 and 6.

Laser CVD can also be used to deposit layers of Si_3N_4 from the gas phase [6.10,32]. In general the physical properties of the films in terms of adhesion, etch rate, stress, stoichiometry, pinhole density, have been found to be comparable to plasma CVD films, and superior to mercury sensitized "photride" films [6.32], especially when photons of 193 nm radiation were present at the surface during the reaction. However there was some indication of slightly more hydrogen contamination in the LCVD films. The electrical properties of the layers, on the other hand, were significantly inferior to conventional CVD films. DEUTSCH et al. [6.10], for example, found using capacitance-voltage (CV) measurements that a large fixed-charge density exists in the LCVD films, which exhibited flat-band voltages in the range 24-30 V, compared with values of 0-1 V for conventional CVD layers. Rather than being a consequence of the generation of traps in the material by the UV radiation, this appeared to be a result of significant contamination by sodium impurities. It seems therefore that the application of these films may be limited to encapsulation, rather than as electrically active layers.

PADMANABHAN and MILLER [6.33] recently employed mercury photosensitized depositions of nitride films from a NH_3+SiH_4 mixture, achieving dielectric breakdown strengths occasionally up to 4 MV/cm. Although the step coverage of the films was always excellent, their refractive index, stoichiometry, and pinhole density were sensitively dependent upon the ammonia/silane ratio and the deposition temperature. The researchers are currently addressing possible special application areas conducive to this essentially "low"-temperature (<800°C) technique, since it avoids sample exposure to excessive charge during processing, unlike other competitive low-temperature technologies such as plasma deposition.

A unique approach to photo-depositing films without sensitizing techniques has been taken by TAMAGAWA et al. [6.34], who used a microwave excited VUV lamp to excite various gas phase species. This apparatus used highly excited deuterium, D_2, as the source of 121.5 nm (10.2 eV) photons, which not only dissociated the reactants, SiH_6 and NH_3, but which were sufficiently energetic to ionize them. Additional ionization of the dissociated species as well as the substrate was also possible with this method. Results showed that the 1.3 nm/min nitride growth obtained solely by the photo CVD route on substrates at 320ºC should be increased to 10 nm/min by incorporating ionization effects. By comparing the etching rates of variously prepared nitride layers, it was found that the VUV grown films were more dense than those grown by Hg lamp or excimer laser, and considerably more dense than plasma enhanced CVD films [6.34].

In addition to growing SiO_2 films by CW CO_2 or Ar ion laser pyrolysis, SUGIMURA and HANABUSA [6.31] also deposited spots of silicon nitride from mixtures of SiH_4 and NH_3. The Arrhenius characteristics of the reaction provided ideal conditions for the production of various mound-like shapes to be used as microlenses. Using this technique, lenses of about 0.16 mm in diameter and focal length of 0.6 mm were manufactured.

High-power CO_2 lasers have been used to produce silicon nitride films by laser-assisted evaporation in the same manner as recently described for oxide formation [6.35], and also by irradiating a mixture of reactive gases, to form ultrafine powders by homogeneous nucleation [6.36]. The latter technique can form the basis of a novel coating technology for use in the electronics industry. Both YARDLEY et al. [6.36] and KIZAKI et al. [6.37] have synthesized fine, high purity Si_3N_4 powders using a CO_2 laser to initiate nucleation in NH_3/SiH_4 flowing gas mixtures. Either species may possibly have been vibrationally excited by the radiation from the laser, according to a wavelength study performed by KIZAKI et al. (Fig.6.6). This figure shows a peak reactivity at a wavelength of 10.71 μm, despite

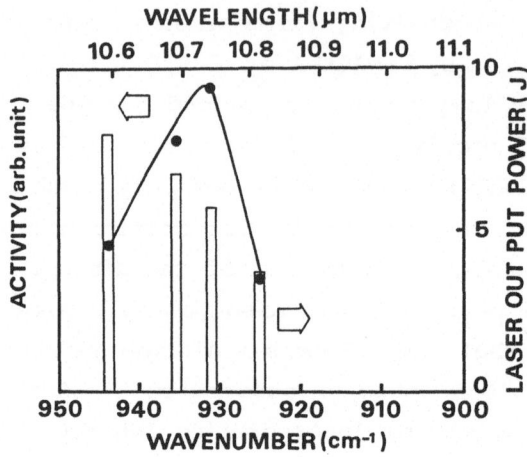

Fig. 6.6. Relationship between the amount of silicon nitride powder deposited (defined here as "activity") against laser output power for various wavelengths [6.37]

the incident laser power not being highest at this point. It is worth mentioning here that the excellent purity of the powders produced by this method is superior to that obtained with conventional techniques. Any trace amounts of Si (and consequently SiO_2 during handling) could be reduced by heating in N_2 gas at 1000°C, resulting in an almost pure white powder. Progress in these areas, which will doubtless include the development of other types of useful compounds, should be worth following in the coming years.

6.5 Organic Polymer Formation

To conclude this section on laser-assisted formation of insulators, we shall discuss the area of direct patterning of organic materials, in particular photoresist layers. The popular traditional method of relief definition in photoresist material is a most important and complex process. First one must coat the substrate with an extremely uniform layer of photoresist, bake it, expose it through an expensive mask to ultraviolet radiation, develop it, post-bake it, and finally etch off the unwanted and strictly defined regions to expose the underlying material in preparation for the next active process in the production line. This sequence is religiously repeated every time a new type of material has to be patterned into the microcircuits.

Laser applications in this area have been centred around the bulk deposition of the photoresist material followed by a laser-based definition technique, such as ablation, or etching, either by direct writing, or by the projection printing technique. These methods will be discussed later in Chap.7.

The relatively small number of reports of photodeposition of patterned organic polymer have the common feature of their use of catalysts in the process. The earliest example employed a surface catalysed reaction in which adsorbed layers of the small organic molecule methyl methacrylate (MMA) were polymerized into distinct regions of polymethyl methacrylate (PMMA).

The incident ultraviolet laser radiation is absorbed by the molecular layers formed on the sample surface as a result of exposure to the MMA atmosphere, thereby creating a free-radical-catalysed polymerization, which is continually refuelled by the continuing collisions of the vapour phase molecules [6.38]. Film thicknesses of typically 0.5 μm have been successfully patterned at net deposition rates of approximately 40 nm/s.

This rate of polymer production, however, can be substantially increased by various sensitization techniques. For example, an intermediate chemisorbed layer of an organometallic compound was found to enhance the growth rate of these films by nearly two orders of magnitude under otherwise identical conditions. Specifically, pre-exposure of the SiO_2 substrates to $Cd(CH_3)_2$ led to PMMA thicknesses exceeding 1 μm [6.39]. This two-step process has also been used to polymerize hydrocarbon monomers, such as ethylene or acetylene [6.40]. In these cases the first step was the controlled exposure of the substrates to UV radiation and a mixture of $TiCl_4$ and $Al_2(CH_3)_6$ vapours. When these molecules undergo co-adsorption into a two-component adlayer, an efficient reaction can be induced by the same UV radiation, which results in the formation of a crystalline compound film. Then, in the second step, in the absence of any UV

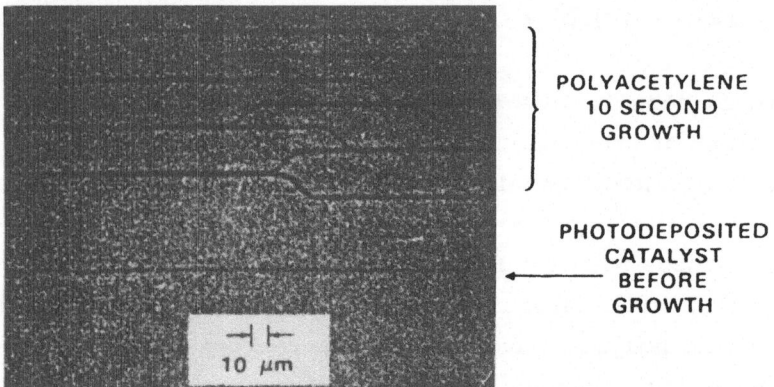

Fig.6.7. Polyacetylene layer produced on approximately one monolayer of photodeposited catalyst. A catalyst film of several monolayers average thickness without polymer growth is also shown for comparison [6.40]

radiation, an organic monomer introduced at a pressure of 100–700 Torr subsequently polymerizes at room temperature on the pre-deposited regions of catalyst material. Fig.6.7 shows a pattern of polyacetylene film formed in this way. The linewidths are well defined by the 3 μm UV laser beam.

Although these experiments appear to offer some new possibilities for direct discretionary patterning on semiconductor microcircuits, the lack of interest in the area is due to the fact that this alternative to conventional processing still requires the intermediate medium of a photoresist in order to define the specific regions to be modified. The more attractive and more far-reaching step is the further possibility of patterning and processing the appropriate material directly with the laser beam, thereby dispensing with the use of photoresist. This is why there is the enormous interest in direct writing, or laser pantography, for producing patterned oxides, nitrides, polycrystals, silicides, metals, and locally doped regions within the device. Nevertheless, this by no means implies the abandonment of polymer deposition techniques, any more than one should dispense with current pattern definition technologies. One cannot ignore, however, the significant possibilities recently opened up through the development of electron-beam writing of specially designed layers. These new techniques may enable further developments of circuit design and layout, and even the possibility of totally novel structures, such as three dimensional device configurations.

7. Material Removal

This chapter presents an extensive review of virtually all aspects of laser-induced material removal. The various reaction modes covered here are pyrolytic and photochemical etching, photon-induced ablation, controlled evaporation, and reactive cutting. A compilation of the different combinations of lasers and presursors used in recent years to pattern and develop a wide variety of materials, including not only insulators and polymers, but also metals and semiconductors, is presented here, along with the recorded removal rates.

7.1 Introduction and Background

As we have already seen in Chapters 4-6, the laser offers attractive and realistic possibilities for the direct writing of thin microstructures and extended films by inducing deposition, growth and surface transformations. These processes involve the addition of materials to the surface of the workpiece, or the rearrangement of pre-existing constituents into a new compound or structure. In this chapter the laser induced mechanisms that lead to material removal from the sample surface, processes such as scribing, trimming, and etching, will be described. The delineation of surface structures by these methods can be readily understood to be complementary to the deposition processes previously described.

If material removal is extended from a primarily surface-treatment phenomenon to a deep penetration process such that holes or vias are formed, then terms like laser-drilling and laser-cutting become more appropriate. Many of these techniques involve laser-induced heating and evaporation, with strictly very little chemical processing as such. Since these areas have already been the subject of many investigations and subsequent applications over the years, and have also been comprehensibly reviewed in several notable publications [7.1-9], they will not be addressed at length here. Instead the main points within each topic will be briefly summarized, and any relevant or potentially important chemical processes will be highlighted where possible.

Etching, on the other hand, involves by definition some chemical reaction, which is traditionally acid-based. In this chapter the uses of reactive agents other than acids leading to pattern definition will also be covered. The central theme behind the fabrication of microelectronic circuits is the concept of repetitious and systematic transfer of specific patterns on to a wafer in a delicately prearranged manner. The photolithographic tradition of relief definition in these cases relies upon an information projection technique which ultimately forms the basis for promoting strictly localized chemical etching for material removal. Film deposition and growth, or localized doping, can subsequently be confined to these newly exposed surface regions and in this way predetermined structures can be prepared.

It is this fundamental and traditional circuit-forming technique that laser direct writing hopes to replace. Laser-assisted etching offers a complementary approach to laser direct writing such that the complete scheme of the pantographic ideology can be fulfilled. In this way, a blanket coverage of some material onto the substrate, by whatever method, can be followed closely by localized and selective removal through photochemical or photothermal mechanisms, thereby leading to an exact representation of some predetermined structural relief.

237

Conventional etching techniques, involving the immersion of batches of wafers in solution, the so-called "wet-chemical" process, are under strong pressure from the uncompromising trend towards dimensional reduction, probably more so than any of the other production methods currently employed. It is not only the inevitable increase in circuit complexity but also the reaction uniformity which will become more and more crucial as production wafer diameters increase towards and above the 150 mm (6 inch) mark, that will demand significant improvements in etching fidelity. Probably the most serious disadvantage of the present-day wet-chemical etching is the phenomenon of under-cutting. This describes the lateral etching that occurs simultaneously with depth definition, thereby severely reducing the spatial resolution defined by the photoresist. It is nonetheless expected, that there will always be requirements for mass batch etching in the less critical lower resolution applications such as surface precleaning, anisotropic etching applications, or various research-based needs such as defect investigations. Here, since a large range of cheap and very selective etches has already been established, alternatives may not be necessary.

By far the most common approach to improving etch resolution has been through the application of reactive gaseous plasmas, a trend towards "dry etching". Plasma etching can in fact routinely define features whose dimensions approach the submicron scale. There are, nevertheless, some drawbacks associated with this novel technology. There are limitations on the ability to create the appropriately reactive compounds for all the applications required. There is also the serious side-effect of radiation damage to some layers. For device dimensions in the future pushing below 0.5 μm, these damage features will become the subject of much concern. Furthermore, this technology adds yet another degree of complexity and expense to the production line, and a gigantic headache to researchers faced with the problem of understanding the complex fundamental chemical kinetics and also trying to advance and widen the spectrum of reactions available.

Research into laser-assisted dry and wet etching has been very active in recent years. The photoetching technique is conceptionally very simple and consists of a few basic steps. Firstly, a previously inert or relatively inactive species becomes chemically reactive near the sample surface as a result of one (or more) of the many laser-induced phenomena discussed in Chap.2. These can include direct photodissociation, pyrolytic decomposition, adsorption, catalytic surface chemistry, or some form of surface excitation. The creation of these necessarily reactive atoms or free radicals in the proximity of the surface can be either in the liquid phase or gas phase, or even in the solid phase itself as a result of adsorption processes. This reactive species subsequently interacts with the atoms of the substrate to form a different compound which can then be preferentially removed from the surface, ideally during the irradiation. The basic mechanisms behind these various processes have already been treated in Chap.2.

In addition to these more common and accepted etching reactions which are determined by the interplay of the chemical species under the appropriate environmental conditions, a novel mechanism for laser-assisted pattern definition by material removal has been suggested in recent years, namely laser ablation. The mechanism is triggered by the fact that most organic materials, polymers and biomolecules strongly absorb radiation in the ultraviolet. By applying photons at the appropriate energy and intensity, e.g. as little as 25 mJ/cm^2 for 5 ns duration F_2 excimer laser pulses [7.10], rapid disruption of the bonding is possible before the absorbed energy can be redistributed within the bulk atoms of the material. Although a detailed mechanism of laser ablation has not yet been fully established, the advantages of this technique are clearly that there is negligible heating in the surrounding material and also there is no particular requirement for special assist gas or high vacuum environment.

In this chapter laser assisted etching (and ablation) of a variety of insulating materials will be reviewed, and additionally, the areas of trimming, cutting and drilling will be discussed, in terms of insulator definition and laser-induced chemical reactions.

7.2 Etching

As will be discussed in Sect.7.4, lasers are already being applied in areas where material removal in precise quantities is required at rapid rates and with great accuracy. These well-established methods almost always involve extensive heating and eventual evaporation, and sometimes this can have undesirable side-effects on the remainder of the workpiece. The use of chemical reactions can help to reduce the temperature required for removal of material, and simultaneously enable extremely fine resolution to be achieved. It is for these reasons that laser-assisted etching of a wide range of materials has been intensely investigated over the last few years. As Table 7.1 indicates, laser induced, or enhanced, etching has been performed on many different types of insulating materials, from glass and ceramics, to a wide range of polymers, using almost the complete range of lasers currently available.

The etching reaction, whereby specifically chosen regions of a solid react with (and are eventually removed into) a surrounding gas or liquid, can be initiated or enhanced by a laser beam that is absorbed by either, or all, of the phases. The photons, as in laser deposition described earlier in Chap.6, can serve to promote photochemical changes in the system, or to elevate the thermal activity of the species involved. Generally, when the radiation is absorbed by gas phase molecules, photodissociation occurs, and free radicals are created that subsequently react vigorously with the nearby surface. In this

240

Table 7.1. Laser-assisted etching of insulating films

SOLID	PRECURSOR /AMBIENT	LASER	ETCH RATE YIELD/PULSE	REF.
SiO_2	Cl_2	Ar	0.3 nm/s	7.10
	CF_3Br, CDF_3	pCO_2	0.03 nm/pulse	7.17,16
	HF	CW CO_2	50 nm/s	7.29
	NF_3,HCl, Cl_2,HF,H_2, CF_3Br	CW CO_2	<175 μm/s	7.105
	$:CF_2$,$\cdot CF_3$	KrF	$\leq$0.013 nm/pulse	7.14,15
	H_2	ArF	$\leq$140 nm/pulse	7.19
SiO_2 (film)	NF_3+H_2	ArF	0.07 nm/s (parallel)	7.20
			0.12 nm/s (normal)	
SiO_2 (bulk)			0.13 nm/s (normal)	
Pyrex	H_2	ArF	$\leq$150 nm/pulse	7.28
Glass	CF_2Br_2	ArF,KrF	0.05 nm/pulse	7.13
Al_2O_3/TiC	KOH	Ar	200 μm/s	7.32
	SF_6	Nd:YAG	0.1 μm/s	7.60
	CCl_4	Ar	0.05 nm/pulse	7.60
$SrTiO_3$	Air,H_2	Kr	<400 μm/s	7.71
$BaTiO_3$				
C	Cl_2,NO_2,NH_3	ArF	20 nm/pulse	7.103
Iron Garnet	H_3PO_4	Ar-dye	75 nm/s	7.34
PET(Mylar)	Air, O_2	Ar	10-50 μm/s	7.60
	Air	ArF	<45 nm/pulse	7.109
PET(Mylar) Polymide, PMMA Polycarbonate	Air	ArF	$\leq$0.12 μm/pulse	7.41,42
Nitrocellulose	Air	ArF	0.12 μm/pulse	7.72,44
	Vacuum	F_2	0.1 μm/pulse	7.10
PET(Melinex S) Polymide	Air	XeCl	$\leq$1 μm/pulse	7.43
Kapton polymide Diazona- phthoquinone (Novolak)	Air	ArF		7.47
Polymide	Air,Vacuum	KrF,XeF XeCl	150 nm/pulse	7.49
PMIPK Polyester Polycarbonate		ArF		7.46

way, for example, Cl_2 gas can be photodissociated by argon laser radiation and then used to etch SiO_2 substrates [7.11,12]. Similar reactions are proposed for SiO_2 and CF_2Br_2 [7.13], CCl_2F_2, and COF_2 [7.14,15] and several others, using various excimer laser wavelengths. It is also possible to use infrared radiation, for example from the CO_2 laser, to photostimulate certain species into etching reactions with these solids. In these cases the excitation may involve vibrational activation or multiple photon dissociation of the gas phase species. Such reactions include SiO_2+CDF_3 [7.16] and SiO_2+CF_3Br [7.17].

Once these reactive species have been created in the gas phase they are free to undergo diffusion, free radical reactions, and collisional de-excitation, and as a consequence, much of the intended spatial confinement of the reaction is lost, leading to a dramatic reduction in resolution. One instance of isotropic diffusion out of the laser excited region is discussed by CHUANG [7.12]. Although the argon laser used in this experiment was focused on to the SiO_2 to a spot size of 5 μm, the random reaction of the diffused species with the surface nearby resulted in an etched feature more than 50 μm in width.

In order to achieve higher resolution for these reactions, it is necessary to introduce additional gas phase species into the reaction chamber that act as buffers or can scavenge the free radicals that stray out of the zone defined by the laser beam. For example, CHUANG [7.18] has suggested that C_2H_6 can react with a Cl atom to form C_2H_5Cl and HCl by a single collision. Similarly, CH_4 and CF_3I can very effectively react with F atoms. Since in these cases the concentration of free radicals would be somewhat reduced in the attempt to confine the reaction, the overall reaction rate might be expected to decrease slightly. A further decrease in the rate might also occur if the scavenging gas or the new products resulting from scaveng-

242

ing interfere with the interaction of the free radicals and the solid surface.

While Cl_2 can be photodissociated by visible radiation (the etching rate for silicon can, in fact, be enhanced by a factor of 20 by changing the incident wavelength of the radiation from 514.5 to 457.9 nm [7.19]), HCl requires wavelengths shorter than about 280 nm to be dissociated by photonic means. In this case, etching can only be carried out if the temperature of the substrate is increased significantly. Thus, since silicon absorbs all visible radiation efficiently, pyrolytic etching can be carried out in silicon, but not on SiO_2, which is transparent to these wavelengths.

With the use of the projection system described previously in Chap.3, LOPER and TABAT [7.14,15] have studied a wide range of laser-induced etching schemes, using ArF and KrF excimer lasers. The absorption spectra of several of the precursors used in this study are shown in Fig.2.3. The range of precursors was chosen to generate the $\cdot CF_3$, $:CF_2$ and atomic fluorine species, the latter in particular for etching polycrystalline silicon and selected metals.

From an examination of the one photon absorption spectra, and consideration of the various bond dissociation energies, LOPER and TABAT [7.14] suggested the following mechanisms for the production of the necessary radicals, as summarized in Table 7.2. It was found under the experimental conditions that photolytic processes designed to produce $\cdot CF_3$ radicals etched SiO_2 very inefficiently, while the generation of $:CF_2$ radicals was most successful in etching the material.

In the case of hexafluoroacetone, because of the poor absorption at 248 nm, a collisional photosensitizing reaction was also used to attempt to improve etching efficiency. In the other reactions generating $\cdot CF_3$, i.e.

$$CF_3I + h\nu(248 \text{ nm}) \rightarrow \cdot CF_3 + \cdot I \, , \qquad (7.1)$$

$$CF_3Br + h\nu(193 \text{ nm}) \rightarrow \cdot CF_3 + \cdot Br \qquad (7.2)$$

Table 7.2. Several possible mechanisms by which etchant radicals could be formed, as suggested by LOPER and TABAT [7.18]

RADICAL	PRECURSOR	PHOTODISSOCIATION PROCESS	
$\cdot CF_3$	CF_3CCF_3	one photon	KrF(248 nm)
	CF_3I	one photon	KrF(248 nm)
	CF_3Br	one photon	ArF(193 nm)
	CF_3NO	one photon	ArF(193 nm)
$:CF_2$	CCl_2F_2	one photon	ArF(193 nm)
	CCl_2F_2	two photon	KrF(248 nm)
	COF_2	one photon	ArF(193 nm)
$\cdot F$	CCl_2F_2	one photon	ArF(193 nm)
	COF_2	one photon	ArF(193 nm)

hydrogen was added to the photolysis mixtures in order to scavenge the atomic iodine and bromine radicals which may have produced nonvolatile silicon iodides and bromides on the oxide surface. However, none of these efforts produced effective etching. The final reaction of

$$CF_3NO + h\nu(193 \text{ nm}) \rightarrow \cdot CF_3 + \cdot NO \qquad (7.3)$$

which produces the nitric oxide radical should not form any non-volatile compounds with silicon that could prevent the oxide from being etched. It was therefore concluded that the $\cdot CF_3$ radical is not capable of efficiently etching SiO_2.

Some of the early work in this area, depicted in Fig.7.1, shows a cross-sectional profile of a layer of SiO_2 after etching by 5100 KrF laser pulses at a fluence of 270 mJ/cm^2, when 100 Torr of dichlorodifluoromethane (CCl_2F_2) was mixed with 660 Torr of helium. The calculated etch rate [7.14] was 0.013 nm/pulse. The photolysis reaction was taken to be

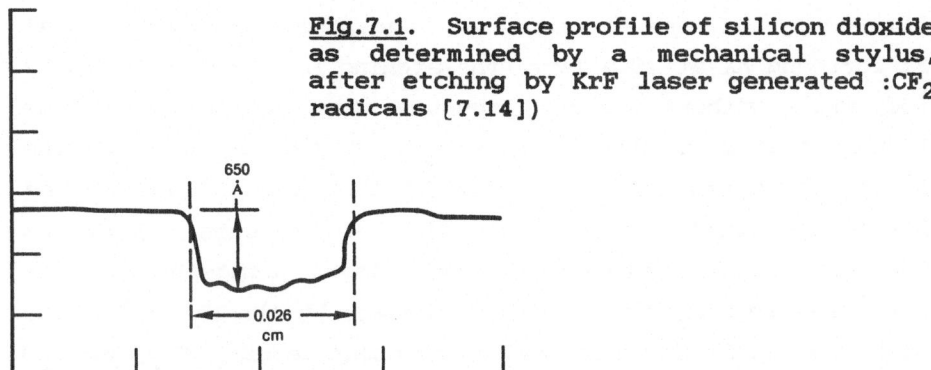

Fig.7.1. Surface profile of silicon dioxide as determined by a mechanical stylus, after etching by KrF laser generated $:CF_2$ radicals [7.14])

650
Å

0.026
cm

$$CCl_2F_2 + 2h\nu(248 \text{ nm}) \rightarrow :CF_2 + Cl_2 \text{ or } 2 \cdot Cl \ . \tag{7.4}$$

It was estimated that removal rates equal to those found in practical plasma etching (200 nm/min) should be possible if this process were extended to cover a 3cm^2 area using an excimer laser capable of producing 200W.

Complimentary work by BRANNON [7.13] on Corning 7070 glass (containing 70% SiO_2, and 25% B_2O_3), showed that CF_2Br_2 irradiated by ArF or KrF excimer lasers could quite efficiently etch the sample at rates around 0.05 nm/pulse. Similar studies with CF_3Br, CF_3I, and SF_6, produced no appreciable etching. It was suggested that the principal photofragment responsible for initiating the material removal was in this case CF_2. In agreement with Loper and Tabat, the generation of CF_3 by this UV photodissociation method did not induce significant etching.

Also using an ArF excimer laser, YOKOYAMA et al. [7.20] have successfully etched SiO_2 by photodissociating NF_3 gas containing 0-4 mol% of H_2. In this reaction, it was suggested that various species such as F, NF, NF_2, and HF, could be created in the gas phase as a result of photolysis of the parent molecule, and adsorb on to the SiO_2 surface to initiate the etching reaction. This was subsequently confirmed by XPS measurements [7.21].

When the SiO_2 is treated with low-pressure mixtures (NF_3 at 3.6 Torr, and H_2 at 0.15 Torr) under irradiation for one minute,

245

followed by the same procedure at atmospheric pressure (NF_3 at 735 Torr and H_2 at 19.5 Torr) the etching is immediately initiated, while without the low pressure step, an incubation time of more than 2 minutes is necessary at the beginning of the high pressure etching. It is thought that the surface of the SiO_2 is chemically activated by the low pressure irradiation step, and that the subsequent processing at atmospheric pressure leads to SiO_2 surface bond cleavage by the N and F atoms which may result in the formation of volatile SiF_4, NO_2, and N_2O molecules, i.e.

$$SiO_2 + NH_3 + H_2 + h\nu(193 \text{ nm}) \rightarrow SiF_4 + NO_2 + N_2O + HF . \quad (7.5)$$

Over longer periods of time, it was suggested that the number of reactive species in the gas phase could accumulate, and this would explain the observed etching rates of 1.3 nm/s for a bulk sample of SiO_2 etched for 40 minutes (Fig.7.2) and 0.12 nm/s for thin layers of thermally grown SiO_2 that could be etched through in less than 7.5 minutes. When the laser beam

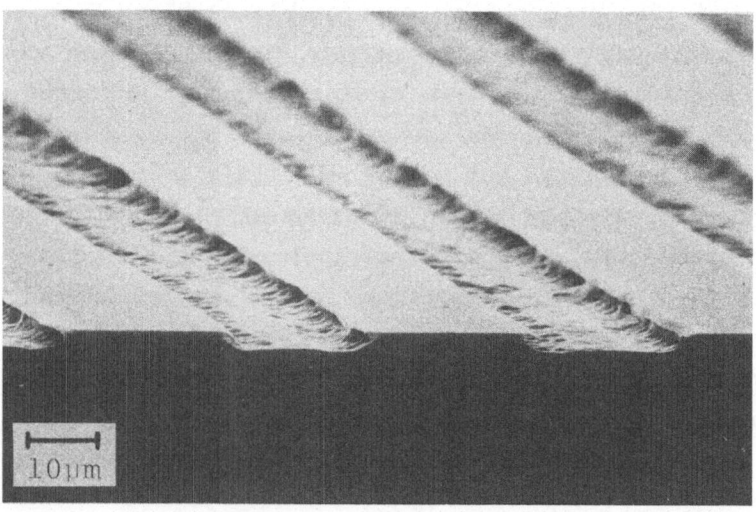

Fig.7.2. SEM photomicrograph of etched silicon dioxide with multiple stripe patterns [7.20]

was parallel to the surface rather than normal to it, the etching rate for the thin films was reduced by almost a factor of 2 to 0.07 nm/s.

There are two possibilities for the observed reduction when parallel illumination is used. Normal illumination could either photostimulate the adsorbate-surface system (Sect.2.4.1), or it could bring about an increase in surface temperature as a result of strong absorption of the laser radiation on the SiO_2 surface. Quartz does not strongly absorb 193 nm radiation as shown in Fig.2.8c and so in bulk samples any difference in the etch rate between the two irradiation geometries would almost certainly be a consequence of adsorbate stimulation (although induced temperature increases may also play a role). In this case the layers were only up to 50 nm thick, and the temperature increase would be strongly dictated by the optical properties of the underlying silicon. The melting threshold for c-Si irradiated by 35 ns pulses at 308 nm is measured to be around 600 mJ/cm^2 [7.22], and has been calculated to be around 400 mJ/cm^2 [7.23]. Other similar experiments give the melting threshold at around 500 mJ/cm^2 [7.24]. By taking the most sensitive of these figures, and even linearly extrapolating back to the 8 mJ/cm^2 used in the experiments, a maximum temperature rise of about 28°C is estimated. There may be a small accumulative effect arising from the repetition rate of the laser pulses used, but it does not seem that thermal arguments can be used in this case to explain the effects observed.

As discussed earlier in Chap.2, dissociation can also occur by multiple photon absorption. This seems to be the dominant mechanism by which SiO_2 has been successfully etched when irradiating CF_3Br [7.17] and CDF_3 [7.16] with a CO_2 laser near the sample surface. Using a parallel irradiation configuration, STEINFELD et al. [7.17] dissociated up to 10% of the CF_3Br molecules via multiple photon dissociation (at 5.5 Torr) presumably creating $\cdot CF_3$ and $\cdot Br$ species that could interact with the SiO_2 surface to form volatile products. Successful etching to a depth of 35 nm at 0.03 nm/pulse with a fluence of 0.4 J/cm^2 was obtained using this technique. This is in contrast to the work of LOPER and TABAT [7.14,15], and BRANNON [7.11], discussed earlier in this section, in which no significant etching was found when photolysing the same molecule with UV laser radiation. It has been estimated from experimental parameters that the average lifetime of the CF$_3$ free radical is of the order of 1.5 μs [7.25] making the effective etch rate 20 μm/s, based on the actual on-time, or duty cycle of the laser. This is at least three orders of magnitude faster than typical plasma etch rates. LOPER and TABAT [7.26] have recently shown that resolution down to the half micron level is achievable in processes involving laser-induced fluorine atom generation. Although this has been reported for Mo and poly-Si, further studies will include other silicon device substrate combinations [7.27].

An alternative method for etching solids that does not involve photolysis but reduces the potential barrier for the interaction of the gas with the substrate, is through direct laser heating of the surface. For example, EHRLICH et al. [7.28] used an ArF laser to thermally excite SiO$_2$ and pyrex in an H$_2$ ambient. Above a threshold of some 200 mJ/cm^2, the strong heating brought about significant material removal in the case of pyrex, which increased with incident fluence up to 150 nm/s and beyond. Rates of up to around 140 nm/s were similarly measured for thin thermally grown silicon dioxide films on silicon. The researchers [7.28] suggested that the more extreme rates of removal were too large to be explained by collisions between the H$_2$ molecules and the sample surface, and concluded that physical or optical ablation could be important in this regime.

The etching process can also be initiated when a gas jet containing HF impinging upon an SiO$_2$ sample is assisted by the simultaneous arrival at the surface of a heating CW CO$_2$-laser beam [7.29]. By raising the substrate temperature in this way, it is expected that the fluorine atoms that have already been adsorbed on the material surface, can more easily interact with the substrate to form the necessary volatile compounds and preferentially etch the SiO_2. Etch rates around 50 nm/s have been achieved by this method.

Although the spatial resolution in the above-mentioned study was of the order of 100 μm, it is possible to improve this when the etching reaction is governed by direct excitation of the surface. This is because the chemical reaction will only occur on the sample where the necessary conditions for the interactions have been satisfied. In situations where the reaction is controlled by the temperature of the substrate, then the resolution will be determined by the laser induced temperature profile, which, as has been pointed out in Sect.2.5, can be within a factor of 2 or less of the beam radius. Furthermore, the larger the activation energy of the chemical reaction, then the more likely it should be for the ultimate resolution to be better than the diffraction limited spot size.

Although our attention thus far has been centred upon material removal in a gaseous phase, it is of course possible to initiate etching in a liquid-phase environment. Indeed, a wide range of methods may be used for obtaining controlled etching in liquids. An early patent by SCHAEFER [7.30] describes the use of photosensitive solution with HCl to etch SiO_2 on Si, using a 1000 W mercury lamp as the source of radiation. An etch rate of about 0.04 nm/s was achieved under those conditions. As the tables in this chapter show, many materials, including metals and semiconductors, have been etched using lasers and liquid etchants such as KOH, H_2SO_4, HNO_3, H_2O_2, H_3PO_4, and $FeCl_3$(aq), but only a few ceramics have been etched in this way. Laser enhanced electroetching (and conversely, electroplating) using simple photosensitive salt solutions such as $CuSO_4$ or $NiCl_2$ as electrolytes has been investigated by VON GUTFELD and coworkers over the years since 1979 [7.31-33].

When melting is induced on the sample surface, the etching reaction rate increases by orders of magnitude. An example

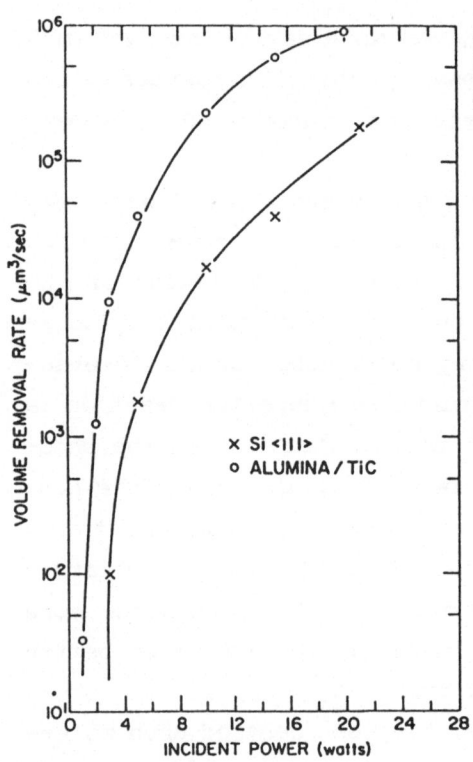

Fig.7.3. Volume removal of alumina/TiC ceramic and <111> silicon, versus laser power [7.25]

x Si <111>
o ALUMINA / TiC

VOLUME REMOVAL RATE (μm^3/sec)

INCIDENT POWER (watts)

taken for Al_2O_3/TiC ceramic immersed in KOH and irradiated with an argon laser beam is shown in Fig.7.3. In this case the potassium hydroxide was 2–18 M, and the incident beam intensity was 10^6–10^7 W/cm². The process has a clear initiation point which has been correlated with the melting point of the material. It has been postulated that for the material removal, the melting and/or evaporation contribute to (a) direct material removal, (b) increasing the surface area in contact with the etchant, and (c) local temperature increase in both the sample and the etchant to promote the thermally activated kinetics [7.25].

Melting of the (111) silicon surface also enables strong etching with KOH to occur. Under standard etching conditions KOH will not etch this crystal orientation although it will etch Si(100) Si very strongly. Melting produces local loss of symmetry of the (111) material thus promoting rapid reactivity. While instantaneous etch rates as high as 200 μm/s have been

250

observed in Alumina/TiC for 1W of incident laser power (10^6 W/cm^2), an average rate of only 15 μm/s has been measured in the etching of through holes in 250 μm silicon with 15W (10^7 W/cm^2) [7.32].

Figure 7.4 shows the morphology of the etched regions of this ceramic material under various conditions [7.32]. Only in the presence of KOH is the wall uniform and the top surface of the periphery smooth and even. Irradiation in air or water produces a variety of less attractive features.

An interesting application of laser induced etching has recently been proposed by ANDO et al. [7.34] for flattening iron garnet films. It was found, for example, that when laser radiation was used to etch these films in phosphoric acid, the etching stopped when the remaining film thickness reached a particular value. It was suggested that in this case the etching rate was proportional to the temperature of the acid, which was heated by convection from the iron garnet film that was directly heated by the laser. A simple calculation showed that the film thickness d_{eq} at which insufficient energy is absorbed by the film itself to maintain a liquid temperature above which etching could proceed, denoted by T_{th} (approximately 125°C), is given by

$$d_{eq} = \frac{1}{\alpha(\lambda)} \ln\left[\frac{1}{1-C(T_{th} - T_0)} \right] \tag{7.6}$$

where $\alpha(\lambda)$ is the wavelength dependent absorption coefficient, C is a constant that depends on the heated area and the heat transfer coefficient, T_0 is room temperature and T_{th} is the threshold temperature. It can therefore be seen that this value does not depend upon the initial film thickness at which the reaction is initiated. This most original technique may of course be usefully applied to a variety of other films and thin film structures.

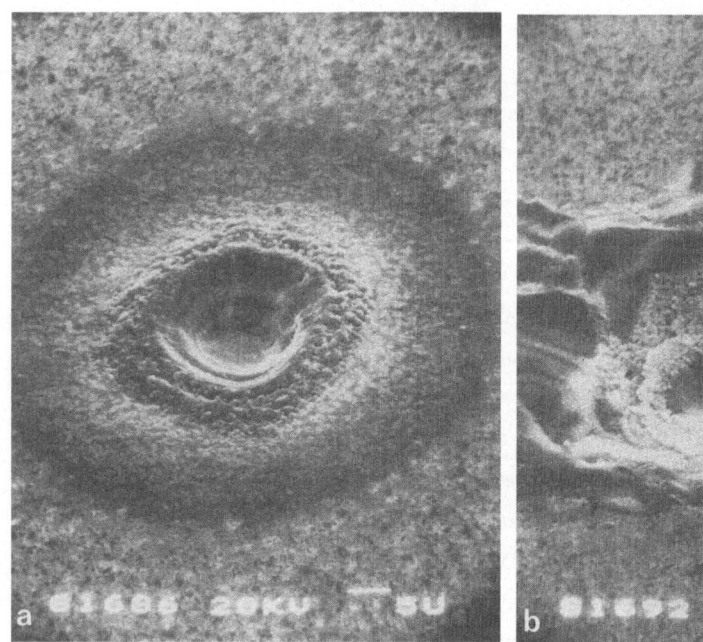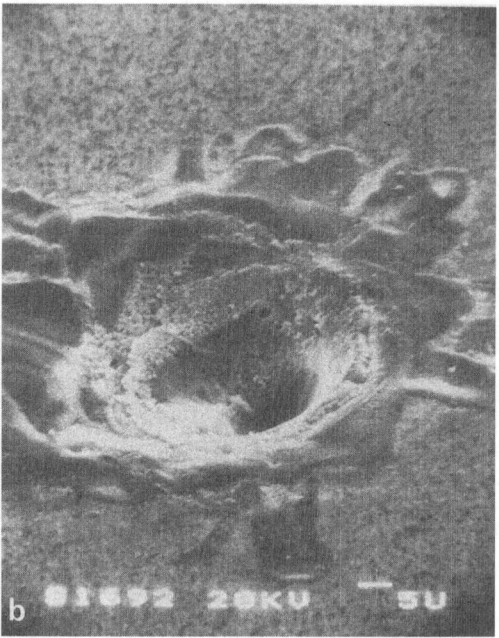

Fig.7.4. Effect of 1.5 W of argon laser radiation for one second on alumina/TiC ceramics in (a) air, (b) water, (c) KOH. Micrograph (d) is a slot etched by scanning the same beam across the sample surface at 16 μm/s [7.32]

7.3 Ablation

The removal of material from a homogeneous surface as a result of laser irradiation can also occur without the use of a reactive intermediate phase. Indeed laser evaporation of materials has been known and applied for many years (see following section, for example). However, when short pulse ultraviolet radiation is applied to specific materials, such as polymers or certain biological tissue, the process of laser ablation may be employed to provide selective removal of the predetermined areas. The ablation process is characterized by its cleanliness, by the extremely sharp definition of the boundaries of the ejected material and that remaining on the surface, and by the complete lack of disturbance to the rest of the layer (see, for example, Fig.7.5), which may be extremely sensitive to even

252

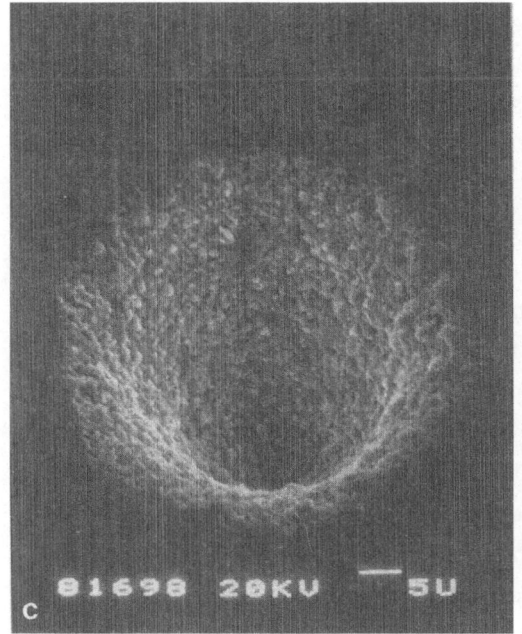

Fig.7.4c

Fig.7.4d

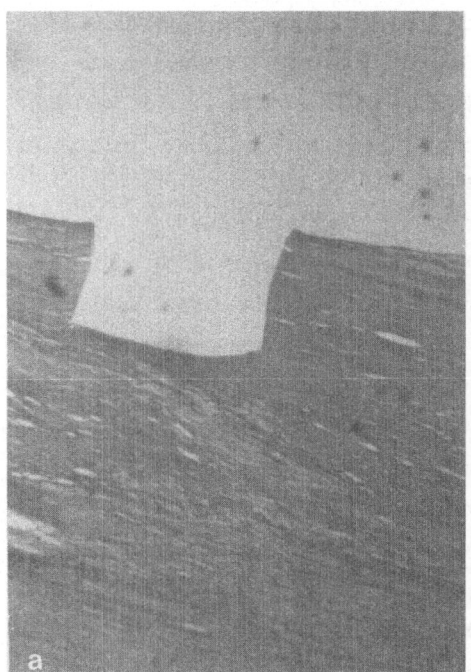

<u>Fig.7.5.</u>   Cross-sections of laser ablated aortic tissue. (a)
1000 ArF laser pulses of 14 ns duration at 2.5 mJ/mm^2, (b) 1800
Nd:YAG laser pulses of 5 ns duration at 10 mJ/mm^2 [7.36,37]

small rises in temperature. The various applications of this technique in the field of photomedicine has recently been reviewed by PARRISH and DEUTSCH [7.35]. Therefore, in this section, only the spontaneous ablation of a range of photoresist materials will be described.

Despite the fact that active layers used in microelectronics, optoelectronics, detector and general heavy industrial technologies are metallic or inorganic in nature, the use of organic compounds in microfabrication is surprisingly extensive. In fact, these compounds form the basis of the lithographic definition of microstructures, acting as photo or electron-beam resists. These are thin films of material that change their structure upon exposure to particular forms of radiation or energy. Usually they are rendered insoluble (soluble) in a developing solution when acting as negative (positive) resists.

A resist material generally contains a polymer, a solvent and a sensitizer. During irradiation of the negative resist polymers, the unsaturated bonds form longer or cross-linked molecules, while in the positive resists the saturated bonds are broken. The sensitizer and the wavelength of the radiation must clearly be chosen to maximize the efficiency of the definition of the ultimate lithographic operation.

In Chap.2 the optical activity of various typical organic groups has already been discussed. The various energy redistribution mechanisms within molecules and solids have also been addressed earlier in the same chapter. For example, we have seen that if sufficient energy can be absorbed directly into a specific bond so as to excite it to a state above its dissociative state, then it will rupture during the following "vibration", which will occur within picoseconds.

We have also discussed a related situation where the absorbed energy is rapidly redistributed around the neighbouring bonds, again very rapidly, but clearly a little slower than the instantaneous bond-breaking event just described. In this instance, one of the weaker bonds in the locality could break, this time through indirect stimulation. These two physically distinct processes are commonly referred to as "ablative photo-decomposition", and "photothermal etching" or "ablation", respectively. It is also clearly possible that in some cases a mixture of these two mechanisms can lead to material ejection.

While the photothermal mechanism is a consequence of an increase in temperature and eventual evaporation, the kinetics behind the proposed photochemical ablation remain to be identified, although evidence of the process has been repeatedly observed [7.36]. This is most strongly demonstrated by the highly heat-sensitive polymers such as atactic polymethyl methacrylate (PMMA) which show no sign of heat damage even when irradiated by pulses as high as 0.5 J/cm^2 that remove the most extremely well defined sections of the film [7.38].

Some of the earlier work using excimer lasers in photolithography [7.39,40] still involved conventional wet processing methods after exposure to the laser beam. However, SRINIVASAN and MAYNE-BANTON [7.41] reported the first use of these lasers to controllably etch the surface of polyethylene terephthalate (PET) films without the need for subsequent processing, i.e. a self-developing process. This could be operated either in air or under vacuum, and etch rates of up to 0.12 μm/pulse were found for 12 ns duration pulses at a fluence of 370 mJ/cm^2. Furthermore, the authors stated that there was little experimental evidence that intense local heating of the film occurred in the irradiation region. SRINIVASAN and BRAREN similarly reported an ablation fluence threshold of only 10 mJ/cm^2 for PMMA films under by ArF laser irradiation at 193 nm [7.42].

Subsequent work by ANDREWS et al. [7.43] also produced ablation of materials such as polyimide and photoresist films that absorbed strongly at 308 nm, including PET. In this case a XeCl excimer laser was used, and etch rates of approximately 0.3 μm/pulse were reported, for approximately identical fluences, although in this case the pulses were 7 ns in duration. By increasing the fluence to around 2 J/cm^2, the rate of removal

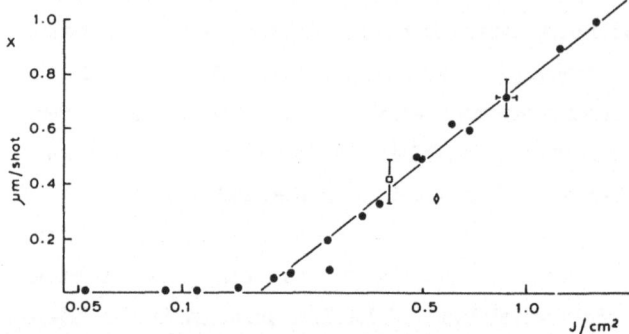

Fig.7.6. Removal rate versus fluence for PET irradiated by XeCl laser pulses, as determined by penetration measurements (·), electron microscopy (), and mass removal ( ) [7.43]

could be increased to more than 1 µm/pulse. Direct etching of polyimide and Shipley AZ1350 photoresists was also achieved in air with fluence thresholds of 60 and 40 mJ/cm^2, respectively.

Figure 7.6 shows the removal rate of PET as a function of incident fluence [7.43]. It can be seen that even at low energy density there is some very small rate of removal. For example, at 50 mJ/cm^2, irradiation of the material at 3Hz for 2 hours removed less than 2 µm from the surface, resulting in a rate of less than 10^{-4} µm/pulse. Progressive discolouring of the film followed by surface charring also occurred under these conditions. Clearly there is a well-defined threshold (around 0.17 J/cm^2) for appreciable material removal. The authors have estimated that the critical temperature in the film at threshold is of the order of 1200 K, and therefore concluded that the mechanism controlling material removal form the surface in this case was rapid evaporation. It was also reported that the morphology of the remaining film was appreciably rough, and that similar structures as well as measurable film removal could be obtained by using a pulsed CO$_2$ laser tuned to one of the vibrational modes of the material [7.43].

The fluence threshold for the ablation of thin nitrocellulose films has been reported to be approximately 20 mJ/cm^2, when using the 193 nm wavelength of the ArF laser [7.44]. Fig.7.7 shows the removal rate of these layers as a space func-

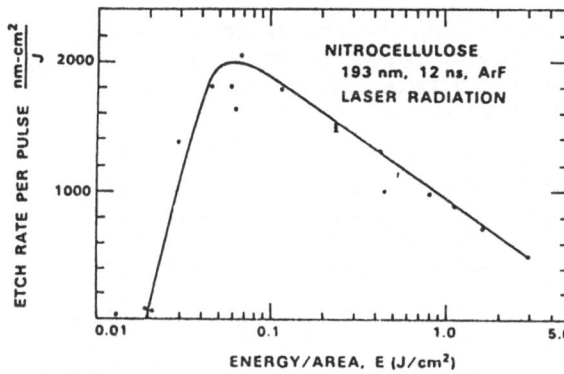

Fig.7.7. Removal rate for nitrocellulose vs fluence for ArF laser pulses [7.45]

tion of incident fluence, and it can be seen that the peak rate is achieved for pulses around 60 mJ/cm^2, at 0.12 μm/shot [7.45]. Direct high resolution photoablation has recently been carried out by RICE and JAIN [7.46] on a wide range of photoresists and polymers (polyester, polycarbonate, poly(methyl isopropenyl ketone) (PMIPK)) and others, using the ArF laser, achieving at times minimum feature sizes of the order of 0.3 μm.

In 1984 LATTA and coworkers [7.47] demonstrated the first use of excimer laser projection ablation as well as the shortest projection photolithography to date by applying ArF laser pulses to films of Kapton polyimide. Their projection system has already been discussed in Chap.3. Polyimide films have, in the past few years, attracted by far the most attention in this area of photoablation (Table 7.3). A number of groups have now reported careful wavelength studies of the photoablative effect in an attempt to isolate any potential photochemical mechanism from the known photothermal effect that unquestionably dominates the reaction at the longer wavelengths.

SRINIVASAN and BRAREN [7.42] showed that the fluence thresholds for removal of polyimide films were 80 and 100 mJ/cm^2, for ablation with the KrF (248 nm) and XeCl (308 nm) lasers, respectively. Subsequently DYER and SIDHU [7.48] reported the values of 30, 31, and 42 mJ/cm^2 for ablation with wavelengths of 193, 248, and 308 nm radiation, respectively, and BRANNON et al. [7.49] gave 27 ± 4, 70 ± 11, and 120 ± 20 mJ/cm^2 for 248, 308

257

Table 7.3. Ablation threshold fluences for some polymer and resist films (using the shorthands, for clarity, λ: wavelength, P.W.: pulse width, R.R.: repetition rate, POLY.: polyimide, N.C.: nitrocellulose, P.C.: polycarbonate)

LASER	λ [nm]	P.W. [ns]	R.R. [Hz]	FLUENCE [mJ/cm²] POLY.	PMMA	PET	N.C.	P.C.	REF.
F_2	157	5	–	–	–	–	25	–	7.10
ArF	193	14	3	–	10	–	–	≃40	7.38,41,42
		11	<3	30	–	28	–	–	7.48
		10	–	–	–	–	20	–	7.44,72
KrF	248	9	<3	31	–	30	–	–	7.48
		15	3	27±4	–	–	–	–	7.49
				80	–	–	–	≃180	7.41,42
XeCl	308	7	10	60	–	170	–	–	7.43
		8	3	42	–	170	–	–	7.48
		35	3	70±11	–	–	–	–	7.49
				100	–	–	–	≃340	7.41,42
XeF	351	15	3	120	–	–	–	–	7.49

and 351 nm radiation, respectively. Meanwhile, KOREN and YEH [7.50,51] measured the fluence at which the surface morphology of the remaining film became smooth rather than rough after ablation, defining this to be F_s. This, of course was not strictly the same threshold T_{th} as the other groups measured, and this resulted in figures of 1500 ± 150, 200 ± 20 and 100 ± 10 mJ/cm² for wavelengths of 351, 248, and 193 nm, respectively. Table 7.4 summarizes this data.

Although the absorption mechanisms for the various wavelengths were often quite different, the trends of ablation behaviour were always very similar in these studies. In particular it was noted by KOREN and YEH that the ArF laser was not unique in producing good film removal, and that other lasers up to a wavelength of 351 nm could perform equally well. By a systematic investigation of the emission spectra during and after irradiation it was found that there were many similarities for the range of wavelengths used. This was interpreted as showing that there was a statistical energy randomization in the films prior to ablation. The process was likened to the regime of the infrared multiphoton dissociation in molecules. Although the pumping mechanisms may be slightly different, the two systems show basically the same features of a partial energy redistribution in the molecules and some nonthermal population distributions once the electronic energy transfer to the vibrational system has occurred in the present process. The differences in the spectra were found mostly in the CN emissions. For 351 nm irradiation, the observations were consistent with complete randomization of the absorbed energy, the dissociation of the weakest bonds in the film, thereby releasing C_2 rather than CN molecules. In the ablation with 193 and 248 nm radiation, a significant amount of CN was present indicating a direct bond breaking process, absorption, and non-thermal localization of this absorbed energy in the CN group, followed by ablation.

KOREN and YEH also found that the product $F_s\alpha$, for each of the wavelengths used was close to a constant value to within about ± 10% (Table 7.4). This was interpreted as showing that the irradiated films were all at the same vibrational tempera-

Table 7.4. Various αF products [kJ/cm^3] for excimer laser ablation of polyimide films; F_{th} is the fluence threshold for ablation, F_s is the fluence threshold for obtaining a smooth surface morphology, and α is the absorption coefficient

| | WAVELENGTH [nm] | | | | REF. |
	193	248	308	351	
αF_s	45	52	—	45	7.50
αF_{th}	3.2	4.3	3.4	—	7.48
αF_{th}	—	4.3±1.5	6.3±2.2	5.6±2	7.49

ture when the rough to smooth morphology transition was observed. However, it is not yet clear precisely what causes this effect on the surface structure.

DYER and SIDHU [7.48] studied the precise fluence threshold values for ablation of polyimide films over a similar wavelength range (193-308 nm). With the knowledge that the optical absorption of this radiation is very strong in these films, and that the thermal diffusion length during a typical pulse duration is less than 50 nm leading to no significant diffusion of heat the system can be described at threshold by one of the special cases outlined earlier in Sect.2.5 as

$$T_{th} - T_0 = F_{th} \ (1-R)\alpha/\rho c_p \eqno(7.7)$$

where T_{th} is the threshold temperature, and T_0 is the starting temperature, and F_{th} is the fluence threshold for material removal. (The other constants are the usual thermal and optical properties of the material). Thus, the product αF_{th} is again expected to be a constant.

This is easily seen from Table 7.4, for a wide range of wavelengths. As is probably the case for laser evaporated silicon discussed earlier [7.52,53], when material is evaporated during the pulse it probably contains a large amount of the absorbed energy, since insufficient time has elapsed for diffusion to appreciably drain it away into the bulk. For the materials described in this section thus far their average thermal diffusivities are of the order of 10^{-3} cm^2/s.

Furthermore, any material that is evaporated during the pulse will additionally shield the surface of the film from the remainder of the incident laser radiation by reflecting and scattering, and of course absorption. T_{th} was calculated to be $\geq$1000 K. The simultaneous study of PET layers led to an even more constant value for F_{th} for ablation of 3.2 ± 5%. It can be seen that this value is very similar to that obtained for the polyimide films.

BRANNON et al.[7.49] have subsequently repeated and confirmed these results (Table 7.4) as well as the interpretations of KOREN and YEH described above. Notice from the table the very close agreement of these data with those of DYER and SIDHU, and the apparent discrepancy with the earlier figures of KOREN et al. It must be remembered, however, that the more recent studies examined the fluence threshold regime, while KOREN et al. defined a special fluence where the surface morphology was seen to change consistently. Nevertheless, the most recent photoacoustic experiments of DYER and SRINIVASAN [7.54], though showing that material is removed after only a short time delay with respect to the pulse, find that the stress transient can reach extremely high levels, and that the temperature rise itself does not appear to be sufficient to account for such rapid ablation. Similar observations of DANIELZIK et al. [7.55] using a quadrupole mass spectrometer, although revealing a thermal Maxwell-Boltzmann velocity distribution of the particles ejected with 193 nm pulses at fluences less than 120 mJ/cm^2, have been stated as "not being in contradiction with the basic ideas of Srinivasan's bond-breaking model ... but point to some additional thermal aspects of the photoablation process". The wealth of evidence strongly suggests the domination of rapid thermal process ablation in the longer wavelength and higher fluence regimes, but it is still not clear how dominant the direct action of bond breaking is at shorter wavelengt before heating takes over.

In the thermal regime, therefore, it appears that the material rapidly ejected from the surface contains the great majority of the absorbed energy, having been removed from the lattice before thermal processes can assist in the diffusion of this excess energy into the remainder of the bulk. However, the precise mechanism controlling the "ablative photodecomposition" in terms of the microscopic nature of the bond-breaking has yet to be established. SRINIVASAN et al. [7.56] have recently proposed that the mechanism of material removal may involve an electronically excited state of the bond being broken, rather than the ground state undergoing rupture associated with thermal decomposition or mechanical pressure. Still further work is clearly required in this area in order to reach a full understanding of the fundamental mechanisms of this process.

DYER and SIDHU [7.57] have recently used projection etching to investigate the characteristics of laser micromachining of free-standing PET films. They have successfully drawn high quality grid patterns on films as thin as 1.5 μm, with a resolution of better than 10 μm, and suggested that with improved optics resolution around one micrometer should be possible. As mentioned above [7.46], features as small as 0.3 μm have been produced on one micron thick resists and polymer films coated on otherwise bare silicon wafers. The smallest features reported so far with an optical self-developing resist technology have recently been studied by HENDERSON et al. [7.10], using contact lithography. By applying F_2 excimer laser pulses to nitrocellulose films, features as small as 200 nm were produced. In this case the threshold fluence was similar to that reported by DEUTSCH and GEIS [7.44] at 25 mJ/cm^2. Fig.7.8 gives two examples of direct ablation of polymer photoresist. Fig.7.8a is a scanning electron micrograph of one of the first commercial photoresists to be patterned by ablation using direct projection of a mask pattern with an ArF laser [7.25]. A higher resolution line pattern, shown in Fig.7.8b, highlights linewidths around 480 nm wide left by material being ablated in a similar fashion on PMMA [7.58]. Theoretically, provided that resists with suitable nonlinear optical responses can be developed, it should be possible to generate images on the VLSI production lines at resolutions below 150 nm.

Tables 7.5-7 catalogue the extent to which laser assisted etching has been studied over the past few years for other materials not strictly covered in this book. For more details on laser induced etching of semiconductors and metals, the reader

Fig.7.8. (a) Commercial photo-resist patterned by ablative laser photo-decomposition [7.25]. (b) Ablated line pattern on 0.5 μm thick PMMA revealing line images around 480 nm. The darker regions are where resist was ablated; only the central line was in focus in the electron microscope (Electron micrograph courtesy R.A. LAWES, Rutherford Appleton Laboratory, UK)

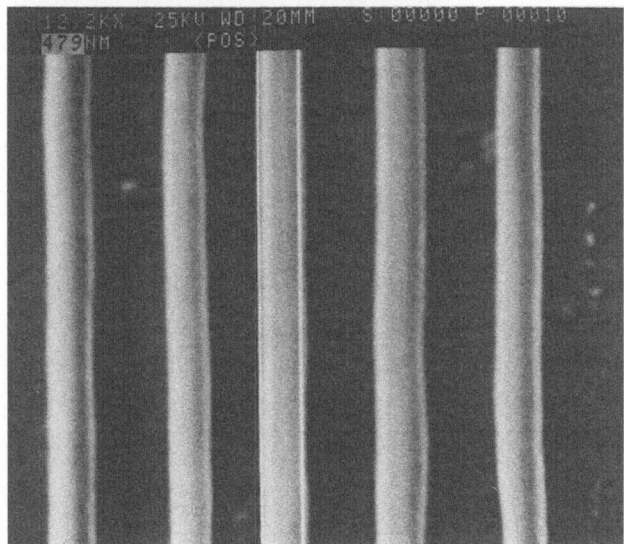

is referred to the review work of CHUANG [7.59,60], and HOULE [7.61].

This section closes with a mention of the approach of LI and OPRYSKO [7.62], which introduces a novel laser-based approach to lithographic technology, making use of a bilayer resist structure. Basically, the top layer consists of a carefully chosen absorbent, which, by absorbing the laser energy, thermally decomposes the deep UV (DUV) dye also present within the coating. The top layer therefore only becomes transparent to

Table 7.5. Laser-assisted etching of metals

SOLID	PRECURSOR /AMBIENT	LASER	ETCH RATE YIELD/PULSE	REF.
Ag	Cl_2 Cl_2	N_2 Nd:YAG	2 nm/s	4.72 7.80
Al	Cl_2 HNO_3+H_2O $H_3PO_4+K_2Cr_2O_7$	XeCl Ar	≥ 1 μm/pulse ≈ 0.5 μm/s	7.82 7.79
Au	Air	Excimer		7.106
Cr	Cl_2	Ar		7.107
MnZn/ Ferrites	CCl_4 CF_4, SF_6 CF_3Cl	Ar	10 μm/s	7.11
Mo	NF_3 Cl_2, NF_3	ArF Ar	≈ 0.022 nm/pulse <3 μm/s	7.14 7.107
$Ni_{0.8}Fe_{0.2}$	SF_6,CF_4 Cl_2,CCl_4	Ar	0.1 μm/s	7.60
Stainless Steel	$NiCl_2$	Ar	10 μm/s	7.81
Ta	SF_6,XeF_2	CO_2	0.26 nm/pulse	7.59
Te	XeF_2	CO_2	1 nm/pulse	7.12
Ti	NF_3	ArF	≈ 0.029 nm/pulse	7.14
W	I_2 Cl_2,NF_3	 Ar	 <1 μm/s	7.104 7.107

Table 7.6. Laser-assisted etching of elemental semiconductors

SOLID	PRECURSOR /AMBIENT	LASER	ETCH RATE YIELD/PULSE	REF.
Si	Br_2	Ar	0.1 nm/s	7.76
Si	Cl_2,HCl	Ar	0.02-10 μm/s	7.19
Si	SF_6	CO_2	0.12 nm/pulse	7.18
Si	XeF_2	CO_2	0.06 nm/pulse	7.73
Si	CF_4/O_2	Ar	8.3 nm/s	7.75
Si	HF(aq)	Ar		7.77
Si,p-Si	Cl_2	XeCl	3.3 nm/s	7.76
Si	KOH	Ar	10 μm/s	7.32
Si	Cl_2	Ar	10 mm/s	7.108
p-Si	COF_2	ArF	0.013 nm/pulse	7.15
Ge	Br_2	XeCl-dye	0.01-0.02 nm/pulse	7.78
Ge	Br_2	Ar	1 nm/s	7.69

Table 7.7. Laser-assisted etching of compound semiconductors

SOLID	PRECURSOR /AMBIENT	LASER	ETCH RATE YIELD/PULSE	REF.
(Al,Ga)As	$H_3PO_4:H_2O_2$	Kr	70 nm/scan	7.88
$GaAs_{1-x}P_x$	Cl atoms	Ar,Ar-dye	0.2 nm/scan	7.101
GaAs	Cl_2	Ar	<33 μm/s	7.87
	CCl_4	Ar	0.3–40 μm/scan	7.86
	Zn	Kr	12 nm/s	7.84
	KOH,HNO_3	Ar	<10 μm/s	7.85
	CF_3Br	ArF	3 nm/s	7.89
	CH_3Br			
	HBr	ArF	(NA)	7.90
	HCl:He	Ar	0.6 nm/s	7.91
	HNO_3	KrF	0.1 nm/pulse	7.93
	HNO_3	Ar	30 nm/s	7.94
		(514,257 nm)		
	CsBr,NaBr,KBr	Kr	100 nm/s	7.94
	CsI,NaI,KI(aq)			
	$H_2SO_4:H_2O_2:H_2O$	Ar,Kr,HeNe	≤10 nm/s	7.96
	KOH,HCl	HeNe		7.97
	$H_2SO_4:H_2O_2:H_2O$			
	$H_2SO_4:H_2O:NaSCN$	HeNe	2.2 nm/s	7.99
	H_2SO_4	Ar	1.8 nm/s	7.62
InP	CH_3Br	Ar	1 nm/s	7.83
	CH_3Cl	(257 nm)		
	CF_3I			
	$FeCl_3(Aq)$	HeNe		7.92
	$HCl:HNO_3:H_2O$	Ar,Kr		7.96
	$H_3PO_4:H_2O$	Ar	20 μm/s	7.100
GaP	$HNO_3:HCl:H_2O$	HeNe		7.92
CdS	$H_2O_2:H_2O$	Ar		7.62
	$KOH:H_2O$			7.96
	KCl,KBr	HeCd		7.98
	$KI:H_2O$			

DUV where the laser has sufficiently heated up the layer. Thus the layer now serves as a positive mask for the underlying coating. This bilayer technique is insensitive to ambient light and for the case of photoresist on PMMA eliminates the usual interfacial problems. It is furthermore a clean technique, and results in features as small as 2 μm in the work reported. It is clear that the laser now offers real opportunities for exploring promising new technologies for improving lithographic processes, for both explicit situations and general purpose applications.

7.4 Trimming

The concept of using laser beams to adjust the resistance of thin-film resistors on completed integrated circuits in situ was first introduced at the end of the 1960s [7.63,64]. Today, it remains one of the most wide-spread laser applications on the production lines of the semiconductor industry. It has found extensive use in tailoring the electrical properties of various geometrical configurations in the film circuit technology commonly known as "hybrids". In the most advanced cases of this technology, circuits are formed by adding discrete miniature components to a network of conductive and resistive tracks formed either by an etching process (in the thin film case) or by screen printing (in the case of thick film circuits). Historically, however, the trimming process has involved no more than localized thermal evaporation of the resistive film. In this section, we shall discuss recent efforts to achieve the necessary resistive tuning by inducing selective chemical reactions.

Figure 7.9 shows schematically typical geometrical pattern of a film circuit including a resistor element. The properties of the resistive material can be modified in such a way that the appropriate degree of conduction required can be fine-tuned from some arbitrary starting point, which, by tradition and for materials used by necessity, is always larger than the value eventually needed.

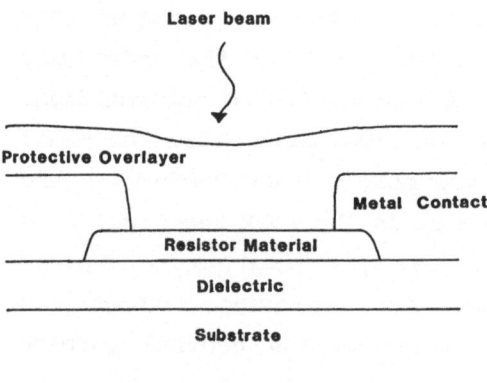

Fig.7.9. Schematic representation of a simple film circuit containing a resistor

The resistance is obviously dependent on the type of material used, the thickness of the layer, and the area between the metal contacts. One conventional method used to alter the resistance involves air-blasting the edges of the resistive material with an abrasive to reduce the overall area thereby increasing the resistance. However, since trimming usually takes place towards the end of the manufacturing line, the circuits can be quite densely packed particularly in recent years as dimensions have continued to shrink, and therefore the advantages of localized processing afforded by laser technology become more desirable, and the increased costs of installing the more expensive equipment can now be justified.

There are three main methods which utilize lasers to adjust the values of the resistive layers. The first is analogous to traditional abrasive blasting and involves controllable evaporation of material in order to alter the geometry of the structure, the so-called edge-shaving technique. The second method also involves material evaporation, but in this case fine sections are cut out of the resistor layer thereby changing slightly the overall geometry and electrical conduction. These can be of almost any shape; straight lines are most common and easiest to program into the moving apparatus, but serpentine and L-shapes are also possible.

The third method, and the one most relevant to our present interests, relies on the use of the laser to alter the actual chemical composition of the thin layer of resistor material. In this way, a previously very conducting layer can be transformed into an electrically insulating material. Therefore, once again the resistance can be progressively inched towards the desired higher value. One recent study used CrSi as the resistor material [7.65]. Investigations showed that CrSi deposited on SiO_2 contained a significant amount of oxygen, mostly in the form of SiO_2, and that upon irradiation the Cr diffused away from the heated zone to mix with the adjoining layers. The remaining material, including some detectable Cr was found to oxidize to various states of chromium oxide and SiO_2. In order to show that chemical changes rather than physical material evaporation played the dominant role in this trimming process the experiments were performed with a passivating overlayer of SiO_2.

Probably one of the most desirable aspects of the laser trimming technique is that it is usually applied at the end of a series of production stages, and consequently, it can be used as a final quality control step. Furthermore, the trimming can not only be performed on the passive circuit, but also with the device receiving the appropriate operating signals, so that active real-time trimming can be performed. In such cases, however, one must cautiously allow for heating, and other effects of the direct laser beam during the measurements, but these are not a major problem.

There is no doubt that laser trimming has been a strong factor in the steady progress of thick film circuit technology. Indeed, not only can it approach the performance of some of the more conventional integrated circuit technologies, but it can offer alternative possibilities for the design of multilayered devices that may not be feasible with the more advanced microelectronic fabrication techniques.

7.5 Cutting and Drilling

Laser cutting and drilling of various materials (e.g. metals, cloth and ceramics) has been an accepted production-line facility for many years now [7.1-3]. In most instances, thermal evaporation of material by strong absorption of the laser radiation is the primary mechanism, and little or no chemistry is involved. However, as we shall discuss during this brief review, the composition of the surrounding ambient can often be important during some of these processes, and can even contribute towards laser induced chemical reactions that assist material removal.

The prime objective of these processes is that material be removed from the appropriate region of the sample as efficiently and as cleanly as possible. All of the methods described here are also characterized by the common requirement that material evaporation is required. We have already discussed earlier in the chapter the alternative laser-induced chemical etching approach to achieving essentially the same end result, as well as the ablation process available with short pulse UV radiation on polymeric insulators.

Conventional laser machining has been demonstrated and indeed has become established over the years as a reliable industrial technology because of its ability to perform localized machining processes on a wide variety of materials on a scale as small as microns. In fact, evaporative laser drilling can produce holes of very small dimensions that cannot be manufactured in any other way. Lasers are presently routinely used to make holes in ruby discs for watch bearings, and in plastic plates for screen printing. The method is also being investigated by Westinghouse for providing self-protection for silicon thyristors.

Just as drilling is achieved for a stationary system, cutting can be achieved usually by introducing little more than movement of the workpiece to the incident laser radiation, or vice versa. As in the laser drilling application, material is removed from the sample by evaporation.

Again in these high power applications, the most desirable property of the operation is the ability to localize the heat treatment. The aim of the process is to introduce the appropriate amount of energy into the sample to melt and evaporate the desired quantity of material to achieve the required structure, without unnecessarily heating up the remainder of the sample. Because of this, both thin foils and thick sheets of material can be successfully cut without any buckling or warping, providing the appropriate laser beam power and scanning speed are used.

In laser cutting, gas is virtually always supplied with the radiation in a coaxial or occasionally in an inclined direction. Such gas jets are used to protect the beam steering or focusing optics, to assist in material removal, and also to prevent or to generate particular chemical reactions. If flammable materials are to be cut by the laser radiation, then an inert gas such as nitrogen or argon is usually used as the assist gas. If oxygen is present during the laser heating of these flammable materials then there are evidently problems with spontaneous burning before the evaporation temperature can be reached. In less extreme cases, when evaporation is actually achieved, this chemical reaction is induced at the edges of the cut, and an undesirable charred layer remains on the material near the perimeter. Thus we can see that there are cases where the laser induced reactions must be suppressed in order to obtain optimum processing.

On the other hand, under certain circumstances, laser cutting of particular materials can be beneficially enhanced in the presence of oxygen, rather than an inert gas. In general it has been found that the presence of oxygen as the assist gas can typically increase the cutting rate by at least 50 over the rate in an inert gas. Indeed, oxygen-assisted cutting has been found to be most suitable for reactive metals such as titanium [7.66]. For Ni too, the cutting speed can be increased by an order of magnitude in this way [7.67].

The reason why the presence of oxygen enhances the cutting rate of metals can be understood by noting that the oxidation reaction is very often exothermic. The laser must initially be used to heat up the metal to its ignition temperature. The ensuing reaction between the heated metal and the oxygen liberates a significant amount of heat that can be as much as the energy supplied by the laser itself. Once initiated, the reaction often propagates erratically out of control under the strong influence of its own energy, and very great care must be taken to control the operating parameters in order to eliminate this runaway effect, often called self-burning.

The relevance of the use of this chemical technique in cutting is that much reduced powers are eventually required to evaporate the necessary material to invoke the cutting action. Table 7.8 gives some illustrative examples for titanium cutting with various CO_2 laser beam powers. It can be seen, for example, that when only 240 W are used on a film of thickness 17 mm in the presence of O_2, that a cutting speed of approximately 10 cm/s is achieved. By comparison, when an inert assist gas is used, it is necessary to use 3 kW to cut only 6.4 mm of titanium at a reduced rate of 6 cm/s.

Table 7.8. Illustrative cutting rates for pure titanium of various thicknesses using oxygen assist gas [7.67,70] and helium assist [7.66] (see [7.3])

THICKNESS [mm]	LASER POWER [kW]	CUTTING RATE [cm/s]	GAS
0.51	0.135	25.5	O_2
17.12	0.240	10.2	O_2
6.36	3.00	5.9	He
31.93	3.00	2.1	He
51.09	3.00	0.84	He

The advantages of this oxidation reaction in cutting titanium has resulted in complete production application being established in the aerospace industry, where contoured cutting is also employed [7.66]. Interestingly, it has been found that the use of oxygen as an assist gas in the cutting of aluminium is not the ideal choice. Although, for the same reasons as given above, the cutting rate is increased by more than 60% when using O_2 rather than inert gas, the best quality finished edges are still obtained when oxygen is not present.

8. Summary and Conclusions

Previous chapters have described in detail the various mechanisms by which thin dielectric films can be formed and patterned by laser beams. In this chapter, we discuss and compare the growth rates and physical properties of these layers. Since laser processing can also be extended to semiconducting, conducting, and superconducting films, we also discuss and assess the potential applications of the technology to specific device-oriented fields.

8.1 Properties and Applications

This book has reviewed the new concepts of laser processing of thin films, concentrating on oxidation processes and generally the formation and patterning of insulators. The impetus behind these, and related investigations into laser formation of conducting patterns, is the possibility of integrating in-situ thin-film processing into the microelectronic and optoelectronic industries. Lasers, in conjunction with electron and ion beams as well as plasma and rapid thermal processing, offer the basis of a future reduced-temperature fabrication technology with discrete direct patterning capability. Many other related technologies could also benefit from this processing ability. These include areas involving the machining and production of ceramics, plastics, and piezoelectrics, as well as holographic etching, optical recording, and high-quality microwave circuit manufacturing.

The possible applications of laser processing in the VLSI industry will be discussed in this concluding chapter. The extent

272

to which this has been considered over the past few years by many different groups worldwide, is shown in Fig.8.1. Here, the standard processes comprising the modern integrated circuit production line are shown alongside the areas where lasers have been investigatively employed to reproduce the appropriate equivalent operation. In many instances, of course, the laser does not, and cannot, completely reproduce the precise characteristics required in order to fit neatly into the already extremely specialized ways of the production line. Additionally, although not indicated in the figure, completely new operations have been suggested. These points will be discussed later in Sect.8.2.

McWILLIAMS et al.[8.1] first demonstrated in 1983, that functional n-MOS field effect transistors can be fabricated on a silicon wafer using only laser-induced pyrolytic deposition, doping, and etching reactions, to reproduce the required surface structures. Had the use of surface cleaning of the bare silicon wafers, and the application of some form of laser oxidation been effected, then the true ab initio creation of MOSFET transistors with lasers would have been realized for the first time in the field. A process for direct write deposition of p-doped polysilicon with resistivities as low as 1 $m\Omega\cdot cm$ has also been recently developed by the group at Lawrence Livermore National Laboratory [8.2]. In fact it has been used to pattern 1000 gate logic circuits on CMOS gate arrays in less than 10 min total processing time. Figure 8.2 is an example of a CMOS gate array circuit interconnected with laser deposited silicon.

The relevant physical and electrical characteristics of the processed structures have been discussed wherever possible throughout this book. It is generally known that laser-prepared films exhibit electrical and structural properties that compare favourably with those of conventionally formed layers. Together with the potential advantages of many of these processes in particular applications, such as resolution, lack of physical

Fig.8.1. (a) A typical menu of events for fabricating MOS integrated circuits. (b) A comparison of some of the processes investigated in the past few years that relate directly to MOS manufacturing. The number after each process indicates the section in this book which refers to the work.

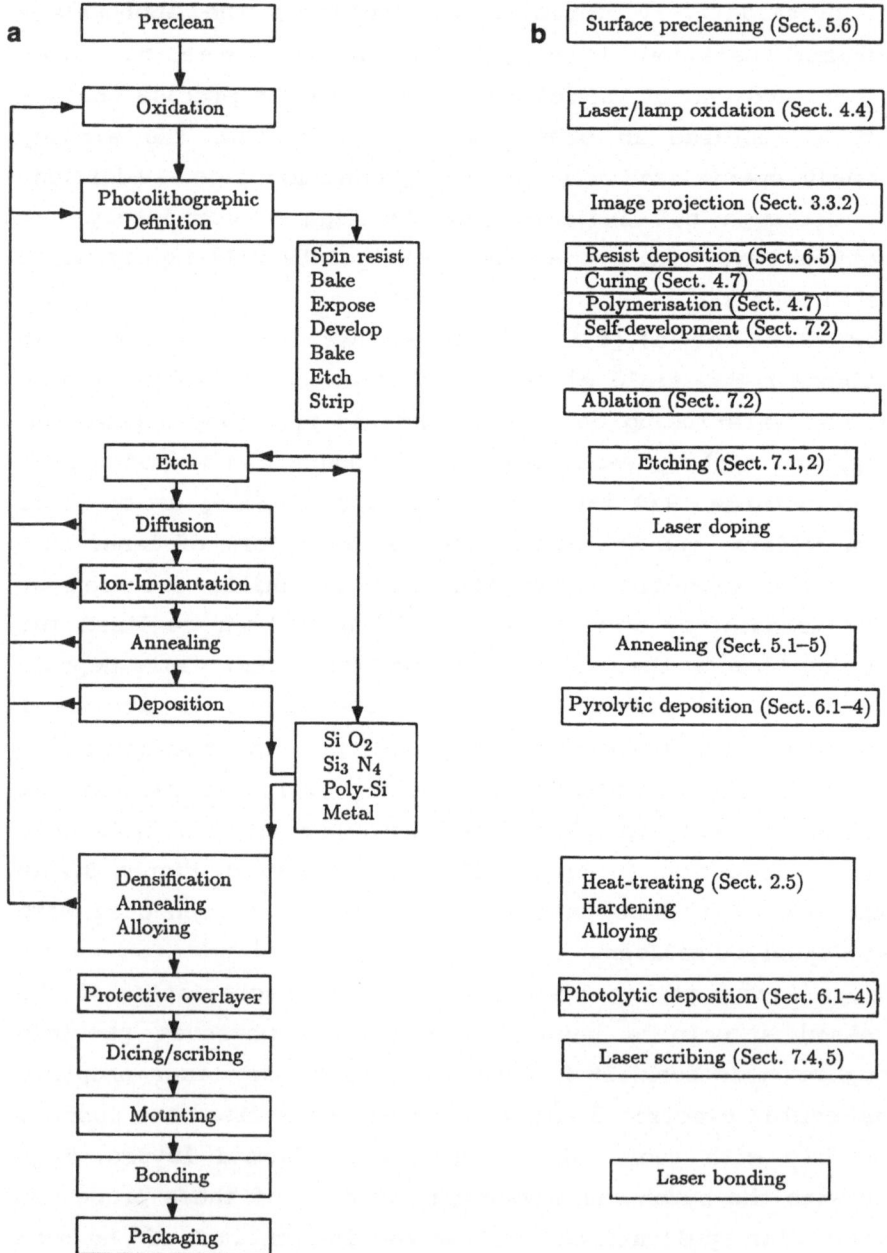

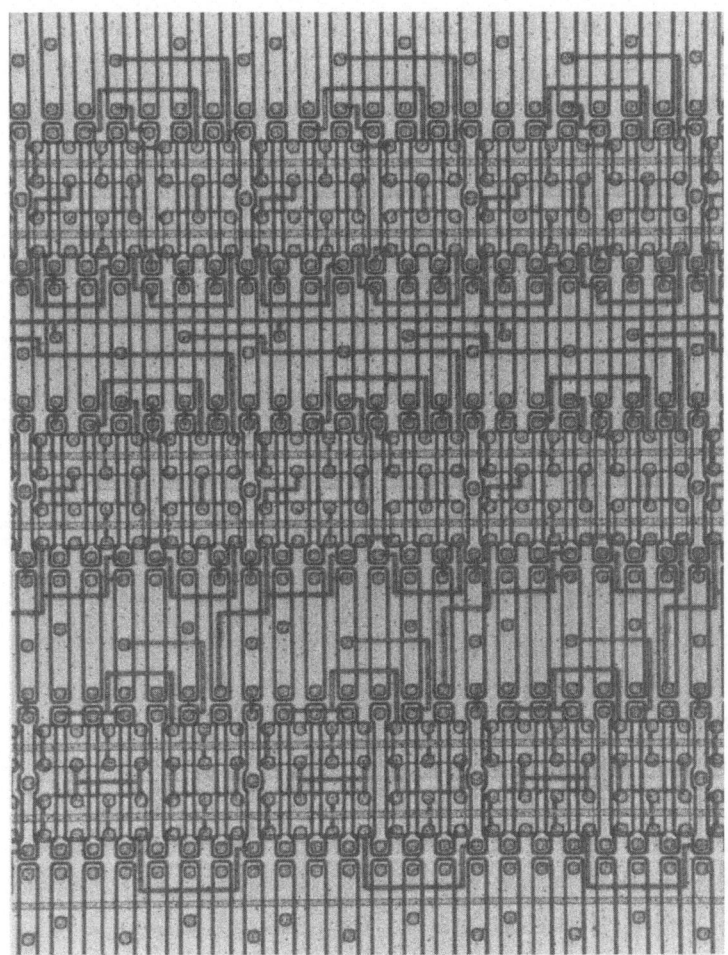

Fig.8.2. Photomicrograph of a CMOS gate array with laser deposited polycrystalline silicon interconnects [8.2]

contact, reduced high temperature cycling, localization, growth rates and chemical efficiency, as well as realistic economic efficiency for low volume preparation, we have seen that these investigations into such a broad range of novel surface modifications more than justify the initial enthusiasm produced by this technology, and the continued investment in this field of research.

An impression of the diversity of techniques that have been employed using lasers to produce thin films can be obtained by glancing through the tables of Chap.4. In order to more completely illustrate the vast differences in the mechanisms controlling film formation by laser beams, details of the growth rates of some thin oxide layers grown on crystalline silicon by a dozen or so different optically-assisted methods have been summarized in Table 8.1, and compared with those found for the more traditional film-forming methods. It can be seen that there is a differential in growth rate of nearly 11 orders of magnitude; there having been several calculations of rates in the region of 10 cm/min, and the report of an astounding growth rate of 6 m/min [8.3]! As indicated in the more comprehensive details of Sect.4.4, however, some of these processes are severely self-limiting once a layer of a particular thickness is grown, usually those which give rise to the more astronomical figures for the rate of growth. Additionally, these layers are often grown under the most extreme thermal conditions induced by rapid melting and solidification. The morphology of these oxides is often very non-uniform and inhomogeneously non-stoichiometric; clearly unsuitable for high quality semiconductor devices.

The growth rate by itself, therefore, is not the sole criterion by which the better film formation techniques are evaluated. One may require, for example, in the case of oxides a precise knowledge of dielectric breakdown strength, bulk and interface charge content, refractive index, density, porosity, etch rate in various acid solutions, hydrogen, carbon or other impurity content, stoichiometry, stress, etc. Only a small fraction of these properties have been investigated for all of the processes shown in Table 8.1, or in the tables in Chap.4 for other insulating films. The reader is referred to the appropriate literature for complete information on any particular film. Here, only the most commonly reported statistic is discussed, that of the dielectric field breakdown strength. This is shown in Fig.8.3.

Table 8.1. Comparison of the various growth rates for oxidation of silicon by conventional methods, and by photonic initiated methods

TECHNIQUE	TYPICAL GROWTH RATE	REF.
KrF laser enhancement	0.07 nm/min	8.35
Excimer laser heating (N_2O)	0.3 nm/min	8.36
Furnace (O_2, 1000°C)	1 nm/min	8.37
Furnace (Steam, 1000°C)	20 nm/min	8.37
CO_2 laser heating	thermal	8.38
Lamp	$\simeq$ thermal	8.11
	> thermal	8.20
Argon laser	> thermal	8.39
RF Plasma	>30 nm/min	8.41
Pyrolytic CVD	>30 nm/min	8.42
Lamp photolytic CVD	>100 nm/min	8.42
XeCl laser photolysis	>300 nm/min	8.12,13
XeCl laser melting	>600 nm/min	8.9
Ruby laser heating	(limited) 6cm/min	8.43
Nd:YAG (266 nm)laser melting	10cm/min	8.44
KrF laser melting	12 cm/min	8.45
Ruby laser melting	6 m/min	8.3

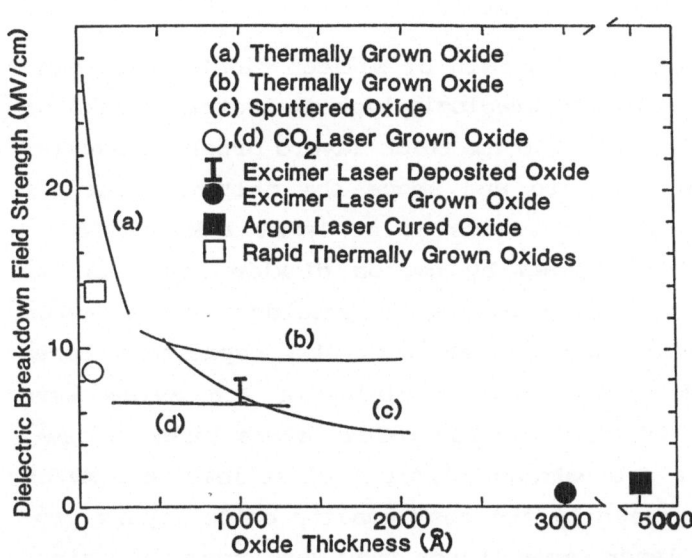

Fig.8.3. Comparison of the dielectric breakdown field strength of various oxide layers grown by conventional methods and by laser-assisted techniques.

Curves (a) and (b) in Fig.8.3 represent some of the highest silicon dioxide breakdown field strengths for thermally grown oxides. [8.4,5]. Curve (c) shows the typical breakdown characteristics for similar films prepared by sputtering [8.6]. The remaining groups of points represent the dielectric strengths reported for a variety of oxide layers obtained by laser [8.7-10], and incoherent lamp processing [8.11]. KRCHNAVEK et al. [8.8] found that 500 nm thick films produced by curing of a spin-on organosilicate with a finely focused argon laser beam (details in Sect.4.7), could sustain electric fields of up to 1.6 MV/cm before breakdown occurred. This compares favourably with layers cured by conventional means.

BOYER et al. [8.12,13], have shown that excimer laser deposited oxides of around 100 nm in thickness exhibit dielectric breakdown strengths between 6 and 8 MV/cm. This group has also reported many of the relevant physical and electrical properties of not only the laser deposited oxide layers, but also of the silicon nitride and aluminium oxide films [8.14]. The oxide properties are found to be very similar to those for RF plasma-deposited films.

Laser-induced pyrolytic growth of silicon dioxide also produces thin oxides with encouragingly high dielectric breakdown strengths [8.7,15]. While 300 nm thick layers grown by excimer laser at fluences close to and above the melting point are known to sustain electric fields of at least 1 MV/cm [8.9], similarly thick films grown by carbon dioxide laser in the solid phase produce field strengths greater than 6 MV/cm [8.7,15]. For films incorporated into MOS capacitors (area approximately 0.2 mm^2) breakdown fields of 6.5 MV/cm were measured, while ultrathin (8.9 nm) oxide layers (area 0.7 μm^2) were found to have a breakdown strength of at least 8.3 MV/cm [8.7]. These are presently the best quality oxide layers produced directly by laser, even though they were grown in a standard research laboratory environment and not within a clean room environment usually used for thermal oxidation.

Because of the present limitations of some laser technology (Sect.8.2), attention has been directed in recent years towards large area rapid photonic heating of silicon through the use of intense incoherent lamps [8.11,16-24]. The application of apparatus comprising lamps such as tungsten-halogen, for rapid thermal annealing, and later alloying, has been studied for several years now. Where the minimization of high temperature processing is desirable, such processing equipment is ideal compared with RF furnaces which have rather slow thermal time constants. Since lamp-heating equipment has been manufactured to semiconductor industry standards of cleanliness, it is therefore not surprising that thin oxide layers with dielectric field strengths as high as 10 MV/cm have been produced [8.11]. For large area isothermal applications, it looks as though the incoherent lamp systems will become extremely desirable pieces of apparatus in the coming years.

8.2 Future Prospects

It is recognized that laser-based techniques could have a impact on sub-micron pattern definition of thin films and devices in areas such as microelectronics, microwave devices and optoelectronics. In this book we have only discussed precise applications in the laser-fabrication of large area and thin film dielectrics, insulators and oxides. The techniques used can of course quite easily be applied to semiconductor and metal films. Whilst many of the vapour deposition techniques have widely been applied to these layers, techniques such as laser evaporation, pulsed plasma deposition, photoablation, and possibly more, have yet to be fully investigated for metals and semiconductors. Novel techniques, however, continue to be examined. For example, significant advances have been made in the new regime of laser direct-write metallization in metallo-organic films at AT&T Bell Laboratories using the technique of oxidation decomposition [8.25,26]. The first report of photoablation of metal films for eventual localized deposition, was only published in the mid-1986 [8.27], at least 6 years after the initial surge of vapour-phase deposition work.

Unlike the rapid integration of laser trimming, bonding, or scribing in the semiconductor industry, the applications of even the most widely studied thin film preparation methods described here are only developing slowly. There are several reasons for this. Firstly, the field is still very much in its infancy in terms of fundamental understanding. As we have seen in Chap.2, it involves the basic concept of photonic interaction with various forms of matter, which subsequently leads to energy redistribution, chemical reactions, and structural transformations, in the gas phase, within solids, and on various surfaces. Other essential mechanisms in the process follow, including adsorption, nucleation and film growth. An immense amount of research is currently being undertaken to more fully explore the mysteries of many of these mechanisms. Once a particular type of reaction pathway has been chosen, and this will be somewhat limited by the availability of well-characterized components, as will be discussed later, significant effort must be invested in optimizing the operating parameters in order to obtain the best quality results.

These techniques were pioneered and have evolved mostly from within specialized research establishments and universities, and only recently has the technologist become fully aware of the great advantages that this laser-based concept could offer. This is clear from the amount of new work in the field which now emanates from the large industrial laboratories in Japan and the USA. From the strengthened practical viewpoints thereby introduced, there is an increasing demand for improvements in the choice, availability, and reliability of components and hardware in the field. But while there exists a lack of fundamental understanding in the basic reaction mechanisms and surface-related phenomena, the requirement for the ideal precursors and lasers is likely to remain somewhat unclear.

Generally, there is a demand for improved UV laser sources and matching gas-phase precursors for low temperature deposition and high resolution processing. The wavelengths where high-power radiation sources give satisfactory characteristics and performance, define a constraint within which precursor development for photodeposition can proceed. For laser pyrolytic growth, the choice of precursor is not so limited, since the technique is akin to conventional CVD, for which many gases have previously been developed. This should not, however, preclude further developement of improved species. In the light of recent successes in the use of laser evaporation techniques [8.27,28] to provide the components for film deposition, of course, the predicted requirement for new precursors may be significantly relaxed.

It is the laser development over the next few years that will probably be the limiting step to major advances in applications in this field. The majority of present-day laser systems are unfortunately not yet suitable for widespread integration into reliable production lines. This is primarily because there has often not been a well-targeted market demand for their application in such areas. While the CO_2 and the Nd:Yag lasers <u>have</u> been given sufficient encouragement over the years for development into satisfactory pieces of high-power industrial apparatus, and the He:Ne laser and laser diodes have acquired low-power credibility through growing reliability, other lasers have not yet achieved similar status. In the ultraviolet, there has not been a significant commercial demand for sources of high power in this region until very recently. Even though significant improvements in performance have been made over the past 18 months in this area, manufacturers still consider that today's research laboratory models are only the forerunners of the true industrial UV sources of tomorrow. The laser-processing advances over the past few years alone are clearly providing a stimulus for more fully investigating the vast range of possibilities opened by this exciting new technology.

It is widely acknowledged that the achievements of present-day integrated circuit production lines are considerable. They are exceedingly successful, but under immense pressure to remain continuously productive. All of the processes incorporated into such lines have been developed to a point of optimization by a deliberate and complex programme of fine-tuning over many years of intensive investigation and employment. Any new process is carefully knitted into the existing production line only when its integration is of obvious and significant benefit to the cost of production, or the yield, or to the performance of the device. It is therefore not difficult to realize that there is massive resistance to the adoption of technologies that would require a significant reorganization of the production-line. So while trimming and scribing can be introduced with relative ease in place of more traditional but less efficient mechanical technologies, processes such as particular aspects of direct writing with lasers cannot be so easily applied without a major impetus from the research laboratories, followed by a complete reorganization of the concept of the production line. As a consequence of this trait, it is foreseeable that in regimes of laser processing where the process can be integrated into existing production lines with only a modest modification to the traditional method, there will be a considerable attraction.

One example of a laser-based method which could be introduced without appreciable disruption to existing production lines, is UV laser assisted image projection, where the laser replaces the incoherent radiation source in modern mask alignment facilities. However, this could be hailed as only the first step towards full acceptance of pattern projection by lasers, where traditional photolithographic operations in delineation could be gradually eliminated. Photo-assisted etching or photo-ablative removal of the resist could eventually be adopted in

place of exposure, development and stripping. Direct patterning
of the photosensitive material by local deposition, for example, would minimize the necessity for the initial spinning and
baking of the resist layer. A more advanced laser application
(only in terms of the number of conventional pattern definition
operations superseded) could eliminate the need for photoresist
material altogether for patterning some layers, by directly and
locally etching the previously deposited or grown layer completely covering the surface. The most significant operation of
all would, within a suitable environment, deposit and pattern a
layer of material directly, according to where the beam impinged upon the surface.

The alternative method to pattern projection involves the
concept of direct writing. One of the biggest attractions of
this mode of processing is that it dispenses with the requirement for masks, which can be expensive and difficult to make.
Here, the beam, or the workpiece, moves according to some pre-
programmed set of instructions, which may be prepared in a similar fashion to that developed for present-day mask design, and
with the appropriate adjustment of the gaseous species present
within the immediate environment, films of the desired properties can be drawn. Since this is instantly seen to be alien to
present-day production concepts, it is clear that such a technology is not going to become established in mass production in
the semiconductor industry. Additionally, a simple calculation
indicates that to draw only one metallization level across a 10
cm (4 inch) wafer at a rapid photolytic deposition rate of 1
μm/s, with a laser beam of approximately 0.5 μm in diameter,
would require almost one month to process each wafer, clearly
an unacceptable processing rate for mass produced devices. Similarly using some of the fastest deposition rates found for
pyrolytic deposition (500 μm/s) gives a more realistic processing time of only a few hours, neglecting any additional time
required for alignment accuracy, directional and angular

changes in movement, or to optimize the actual deposition rate in order to improve the film properties.

Nevertheless, there are limited but well-defined areas where particular aspects of the direct-writing mode of laser processing are already receiving considerable attention, e.g., in photolithographic mask repair, as a result of the initial work of EHRLICH et al. [8.29]. Similarly, laser writing has been applied to the optimization and repair of circuits by several groups [8.30-32]. In such applications, localized deposition of metal can fill in small gaps between broken conducting lines, and small areas of unwanted deposition can be evaporated from the substrate or mask to leave an essentially unblemished surface. Laser writing has also been used for circuit optimization, where the gate of a field-effect transistor can be metallized and linked to neighbouring interconnect lines with a quality that is comparable to that of conventional metallization methods [8.32].

These applications open up a most exciting new possibility for customization of generic circuits, or for improving yields of particularly notorious devices by duplication of problem areas, and linking in the optimum cells when processing is completed. The rapid discretionary interconnection of prefabricated gate arrays, or even programmable logic arrays, is ideal for the development of prototype circuits. The advancement of multilevel interconnectors in recent years could similarly benefit from selective low-temperature removal and deposition of material [8.33]. Elective interconnection of gate arrays has already been reported by EHRLICH et al. [8.34] and McWILLIAMS et al. [8.35]. If real-time interconnection of particular designs can be realized, the possibility of circuit designers being able to draw, correct and modify their plans on an actual circuit may not be too far removed from today's reality. In this way, the turn-around time between error-corrected prototype designs could be accelerated from weeks and months to hours, and testing and debugging phases in chip manufacture could be speeded up significantly.

Where low-temperature processing is essential, such as with many compound semiconductor and multilayered thin-film devices, laser photochemical technology may find useful application. Presently, in the optoelectronics area, production volumes can often be extremely low, and therefore processing time is not a critical parameter. Here, either the direct write mode, or the mask-projection method could be used to form any custom-required device patterns.

The capability of direct writing may also be very desirable in the production of waveguide structures. With the knowledge of the wide-ranging effects that can be induced on the surface of many materials, as described in Chap.2, it would be relatively easy to fabricate graded-index waveguides as well as curved and tapered guiding structures.

Localized laser processing has also been shown recently to be capable of producing vias, or holes, with extremely large aspect ratios. Thus, inter-chip, or inter-board communications could be made much faster by reducing the length of the usual interconnection pathways. Both electrical and optical interconnect technologies could benefit here.

Micromachining and direct writing of a myriad of materials could inevitably bring about many hitherto unrecognized advantages to many less glamorous technological areas, such as in printing, or engraving. It therefore seems to be quite a realistic expectation that in the not too distant future laser processing of thin films will find applications in important low-volume fine-scale processing areas.

Localized laser processing may eventually become an integral part of the perceived requirement over the next few decades for a total in-situ processing capability involving beams of all kinds. Ion, electron, and laser beams, as well as intense coherent lamp systems and plasmas may eventually unite to form a complete technology for fabricating highly customized devices completely within the confined cells of a ultrahigh vacuum system. Under these conditions of low pressures and cold-wall temperatures, external contamination could be virtually eliminated, and a new regime of high quality devices could emerge. Much remains to be done, however, if such dreams are to be ultimately realized.

8.3 Postscript

Further recommendations for additional background information are referenced at the end of this chapter, as well as new reference reviews of interest that have appeared in the literature during publication. In order to keep in touch with future developments of the field in general, the reader is referred to the proceedings of the Materials Research Society meetings which are held annually both in the USA (Boston) and in Europe (Strasbourg).

References

Chapter 1

1.1 See, for example, Lasers & Applications **6** (1) 73 (1987); and Laser Focus **23**, 52 (January 1987)
1.2 J.F. Ready: *Effects of High Power Radiation*, (Academic, New York 1971)
1.3 S.S. Charschan: *Laser in Industry*, (Van Nostrand, New York 1972)
1.4 W.W. Duley: *CO_2 Lasers, Effects and Applications*, (Academic, New York 1976)
1.5 J.F. Ready: *Industrial Applications of Lasers* (Academic, New York 1978)
1.6 J.F. Ready (ed.): *Laser Applications in Materials Processing*, SPIE **198** (1980)
1.7 H.Koebner (ed.): *Industrial Applications of Lasers* (Wiley, Chichester 1984)
1.8 J.F. Ready: Proc. IEEE **70**, 533 (1982)
1.9 *Energy Beam Solid Interactions and Transient Processing*, ed. by D.K. Biegelsen, G. Rozgonyi, C.V. Shank (Materials Research Society, Pittsburgh 1984) and references therein
1.10 J.M. Poate, J.W. Mayer (eds.): *Laser Annealing of Semiconductors* (Academic, New York 1982)
1.11 I.W. Boyd, J.I.B. Wilson: Nature **303**, 481 (1983)
1.12 I.W. Boyd: Contemp. Phys. **24**, 461 (1983)
1.13 D.J. Ehrlich, R.M. Osgood Jr, T.F. Deutsch: IEEE J. QE-16, 1233 (1980)
1.14 T.J. Chuang: J. Vac. Sci. Technol. **21**, 798 (1982)
1.15 D.J. Ehrlich, J.Y. Tsao: J. Vac. Sci. Technol. B1, 969 (1983)
1.16 D.J. Ehrlich, J.Y. Tsao: In *VLSI Electronics: Microstructure Science* **7**, 129 (Academic, New York 1983)
1.17 R.M. Osgood: Ann. Rev. Phys. Chem. **34**, 77 (1983)
1.18 R.M. Osgood, S.R.J. Brueck, H.R. Schlossberg (eds.): *Laser Diagnostics and Photochemical Processing for Semiconductor Devices* (North-Holland, Amsterdam 1983)
1.19 F.R. Aussenegg, A. Leitner, M.E. Lippitsch (eds.): *Surface Studies with Lasers*, Springer Ser. Chem. Phys., Vol.33 (Springer, Berlin, Heidelberg 1983)

1.20 A.W. Johnson, D.J. Ehrlich (eds.): *Laser Controlled Chemical Processing of Surfaces* (Elsevier, New York 1984)

1.21 D. Bäuerle: *Chemical Processing with Lasers*, Springer Ser. Mat. Sci., Vol.1 (Springer, Berlin, Heidelberg 1986)

1.22 M. von Allmen: *Laser Beam Interactions with Materials*, Springer Ser. Mat. Sci., Vol.2 (Springer, Berlin, Heidelberg 1987)

1.23 See, for example, H. Beck, H.-J. Güntherodt (eds.): *Glassy Metals II*, Topics Appl. Phys., Vol.53 (Springer, Berlin, Heidelberg 1983) Chap.8

1.24 Yu A. Bykovskii, A.G. Dudoladov, V.P. Kozlenkov, P.A. Leont'ev: JEPT Lett. **20**, 135 (1974)

1.25 A.A. Agasiev, A.K.H. Zeinally, V.M. Salmanov, V.I. Tagirov, K. Yu Karakurchi: Sov. Phys. Semicond. **9**, 777 (1975)

1.26 I.B. Khaibullin, E.I. Shtyrkov, M.M. Zaripov, M.F. Galyautdinov, R.M. Bajazitov: Paper No. 2061-74, deposited in Viniti, Moscow (1974) (in Russian)

1.27 I.B. Khaibullin, E.I. Shtyrkov, M.M. Zaripov, R.M. Bayzitov, M.F. Galjanudinov: Radiat. Eff. **36**, 225 (1978) and references therein

1.28 T.J. Chuang: Surf. Sci. Rept. **3**, 1 (1983)

1.29 F. Micheli, I.W. Boyd: Optics and Laser Technology **18**, 313 (1986); **19**, 19 and 75 (1987)

1.30 D. Bäuerle (ed.): *Laser Processing and Diagnostics*, Springer Ser. Chem. Phys., Vol.39 (Springer, Berlin, Heidelberg 1984)

1.31 S.D. Allen (ed.): *Laser Assisted Deposition, Etching, and Doping*, SPIE **459** (1984)

1.32 F.A. Houle, T.F. Deutsch, R.M. Osgood (eds.): Laser Chemical Processing of Semiconductor Devices, Extended abstracts of MRS Symp.B (Materials Research Society, Pittsburgh 1984)

1.33 S. Metev (ed.): Laser Assisted Modification and Synthesis of Materials, Sofia University (1985)

1.34 H.H. Gilgen, T. Cacouris, P.S. Shaw, R.R. Krchnavek, R.M. Osgood: Appl. Phys. B**42**, 55 (1987)

1.35 F.A. Houle: Appl. Phys. A**41**, 315 (1986)

1.36 Y. Horiike, N. Hayasaka, M. Sekine, T. Arikado, M. Nakase, H. Okano: Appl. Phys. A**42** (1987) in print

1.37 M.J. Berry: Ann. Rev. Phys. Chem. **26**, 259 (1975)

1.38 P.A. Schultz, As.S. Subd, D.A. Krajnovich, H.S. Kwok, T.Y. Lee: Ann. Rev. Phys. Chem. **30**, 379 (1979)

1.39 K.L. Kompa, S.D. Smith (eds.): *Laser Induced Processes in Molecules*, Springer Ser. Chem. Phys. Vol.6 (Springer, Berlin, Heidelberg 1979)

1.40 A.H. Zewail (ed.): *Advances in Laser Chemistry*, Springer Ser. Chem Phys. Vol.3 (Springer, Berlin, Heidelberg 1978)

1.41 J.I. Steinfeld: *Laser-Induced Chemical Processes* (Plenum, New York 1981) p.243

1.42 J. Hager, Y.R. Shen, H. Walther: In [Ref.1.30, p.154]

1.43 W. Witteman: *The CO_2 Laser*, Springer Ser. Opt. Sci., Vol.53 (Springer, Berlin, Heidelberg 1987)

Chapter 2

2.1 M. Snells, E. Borsella, R. Fantoni, A. Giardini-Guidoni: In *Laser Processing and Diagnostics*, ed. by D. Bäuerle, Springer Ser. Chem. Phys., Vol.39 (Springer, Berlin, Heidelberg 1984) and references therein.

2.2 M. Bass, J.R. Franchi, J. Chem. Phys. **64**, 4417 (1976)

2.3 T.J. Chuang: J. Vac. Sci. Technol. **21**, 798 (1982)

2.4 J.I Steinfeld, T.G. Anderson, C. Reiser, D.R. Denison, L.D. Hartsough, J.R. Hollahan: J. Electrochem. Soc. **127**, 514 (1980)

2.5 T.J. Chuang: J. Chem. Phys. **74**, 1453 (1981)

2.6 F.A. Houle, T.F. Chuang: J. Vac. Sci. Technol. **20**, 790 (1982)

2.7 T.J. Chuang: IBM J. Res. Dev. **26**, 145 (1982)

2.8 R. Solanki, W.H. Ritchie, G.J. Collins: Appl. Phys. Lett. **43**, 454 (1983)

2.9 G.L. Loper, M. Tabat: SPIE **459**, 121 (1984)

2.10 G.E. Jellison Jr, F.A. Modine: Appl. Phys. Lett. **41**, 180 (1982)
     G.E. Jellison Jr, F.A. Modine: Phys. Rev. **B27**, 7466 (1983)

2.11 J.E. Jellison: In [Ref.2.87, p.95]

2.12 N. Bloembergen: IEEE J. **QE-10**, 375 (1974)

2.13 J.A. McKay, J.T. Schriemf: IEEE J. **QE-10**, 2008 (1981)

2.14 R.F. Marks, R.A. Pollak, Ph.Avouris, C.T. Lin, Y.J. Thefaine: J. Chem. Phys. **78**, 4270 (1983) and references therein.

2.15 R.F. Marks, R.A. Pollack, Ph. Avouris: In *Laser Diagnostics and Photochemical Processing for Semiconductor Devices*, ed. by R.M. Osgood, S.R.J. Brueck, H.R. Schlossberg (North-Holland, Amsterdam 1983) p.257

2.16 F.A. Houle, T.F. Deutsch, R.M. Osgood (eds.): Laser Chemical Processing of Semiconductor Devices, Extended abstracts of MRS Symp.B (Materials Research Society, Pittsburgh 1984)

2.17 J. Heidberg, H. Stein, E. Riehl: Phys. Rev. Lett. **49**, 666 (1982)

2.18 T.J. Chuang: J. Chem. Phys. **76**, 3828 (1982)

2.19 T.J. Chuang, H. Seki: Phys. Rev. Lett. **49**, 382 (1982)

2.20 R.S. Lichtman, D. Shapira: CRC Crit. Rev. Solid State Mat. Sci. **8**, 93 (1978)

2.21 N.H. Tolk, M.M. Traum, J.C. Tully, T.E. Madey (eds): *Desorption Induced by Electronic Transitions DIET I*, Springer Ser. Chem. Phys. Vol.24 (Springer, Berlin, Heidelberg 1983)
     W. Brenig, D. Menzel (eds.): *Desorption Induced by Electronic Transitions DIET II*, Springer Ser. Surf. Sci., Vol.4 (Springer, Berlin, Heidelberg 1985)

2.22 D. Menzel: J. Vac. Sci. Technol. **20**, 538 (1982)

2.23 V.S. Antonov, V.S. Letokhov, A.N. Shibanov: Appl. Phys. **25**, 71 (1971)

2.24   D.E. Ramaker: In [2.21]
2.25   B. Schafer, P. Hess: Appl. Phys. B37, 197 (1985)
2.26   D. Menzel, R. Gomer: J. Chem. Phys. 41, 3311 (1964)
2.27   P.A. Redhead: Can. J. Phys. 42, 886 (1984)
2.28   M.S. Slutsky, T.F. George: J. Chem. Phys. 70, 1231 (1979)
2.29   M.S. Dzhidzhoev, A.I. Osipor, V.Ya. Panchenko, V.T. Plato-
       nenko, R.V. Khokhlow, K.V. Shaitan: Sov. Phys. JETP 47, 684
       (1978)
2.30   P.R. Antoniewicz: Phys. Rev. B21, 3811 (1980)
2.31   S. Baidyaroy, W.R. Bottoms, P. Mark: Surf. Sci. 28, 517
       (1971)
2.32   Y. Shapira, S.M. Cox, D. Lichtman: Surf. Sci. 54, 43 (1976)
2.33   N. Van Hieu, D. Lichtman: Surf. Sci. 103, 535 (1981)
2.34   N. Van Hieu, D. Lichtman: J. Vac. Sci. Technol. 18, 49
       (1981)
2.35   T. Kawai, T. Sakata: Chem. Phys. Lett. 69, 33 (1981)
2.36   T.J. Chuang, J. Vac. Sci. Technol., B3, 1408 (1985)
2.37   T.J. Chuang, H. Seki, I. Hussla: Surf. Sci. 158, 525 (1985)
2.38   S.H. Brunauer, P.H. Emmett, E. Teller: J. Am. Chem. Soc.
       60, 309 (1938)
2.39   C.J. Chen, R.M. Osgood: Appl. Phys. A31, 171 (1983)
2.40   D.J. Ehrlich, R.M. Osgood: Chem. Phys. Lett. 79, 381 (1981)
2.41   C.R. Brundle, H. Hopster: J. Vac. Sci. Technol. 18, 663
       (1981)
2.42   F. Bartels, W. Mönch: Surf. Sci. 143, 315 (1984)
2.43   D.J. Ehrlich, J.Y. Tsao: J. Vac. Sci. Technol. B1, 969
       (1983)
2.44   D.J. Ehrlich, J.Y. Tsao: Appl. Phys. Lett. 46, 198 (1985)
2.45   W.Brown: In Laser and Electron Beam Solid Interactions
       and Material Processing, ed. by J.F. Gibbons, L.D. Hess,
       T.W. Sigmon (North Holland, Amsterdam 1981) p.20
2.46   E.J. Yoffa: Phys. Rev. B26, 2415 (1981)
2.47   N. Bloembergen: In Laser-Solid Interactions and Laser
       Processing, ed. by S.D. Ferris, H.J. Leamy, J.M. Poate
       (AIP, New York 1979) p.1
2.48   H.S. Carslaw, J.C. Jaeger: Conduction of Heat in Solids
       (Clarendon, Oxford 1959)
2.49   M. Lax: J. Appl. Phys. 48, 3919 (1977)
2.50   Y.I. Nissim, A. Lietoila, R.B. Gold, J.F. Gibbons: J. Appl.
       Phys. 51, 274 (1980)
2.51   J.E. Moody, R.H. Hendel: J. Appl. Phys. 53, 4364 (1982)
2.52   F. Ferrieu, G. Auvert: J. Appl. Phys. 54, 2646 (1983)
2.53   P. Schvan, R.E. Thomas: J. Appl. Phys. 57, 4378 (1985)
2.54   C.Y. Ho, R.W. Powell, P.E. Lilley: J. Phys. Chem. Ref. Data
       Suppl. 3, I-588 (1974)
2.55   P.D. Maycock: Solid State Electron. 10, 161 (1967)
2.56   E.g. see W.B. Joyce: Solid State Electron. 18, 321 (1975)
2.57   H.W. Lo, A. Compaan: J. Appl. Phys. 51, 1565 (1980)
2.58   I.W. Boyd: Appl. Phys. Lett. 42, 728 (1983)
2.59   I.W. Boyd, T.D. Binnie, J.I.B. Wilson, M.J. Colles: J. Appl.
       Phys. 55, 3061 (1984)

2.60   A.E. Bell: RCA Rev. **40**, 295 (1979)
2.61   P. Baeri, S.U. Campisano, G. Foti, E. Rimini: J. Appl. Phys. **50**, 788 (1979)
2.62   J.R. Meyer, M.R. Kruer, F.J. Bartoli: J. Appl. Phys. **51**, 5513 (1980)
2.63   M. Bertolotti, C. Sibilia: IEEE J. QE–17, 1980 (1981)
2.64   R.F. Wood: Phys. Rev. B25, 286 (1982) and references therein
2.65   D.L. Kwong, D.M. Kim: J. Appl. Phys. **54**, 366 (1983)
2.66   D. Agassi: J. Appl. Phys. **55**, 4376 (1984)
2.67   M.O. Thompson, G.J. Galvin, J.M. Mayer, P.S. Peercy, J.M. Poate, D.C. Jacobson, A.G. Cullis, N.G. Chew: Phys. Rev. Lett. **52**, 2360 (1984)
2.68   I.W. Boyd, T.F. Boggess, A.L. Smirl, S.C. Moss: Optica Acta **33**, 527 (1986)
2.69   I.W. Boyd, S.C. Moss, T.F. Boggess, A.L. Smirl: Appl. Phys. Lett. **45**, 80 (1984) and references therein
2.70   J.F. Ready: *Industrial Applications of Lasers* (Academic, New York 1978)
2.71   See e.g. A.G. Cullis: Rep. Prog. Phys. **48**, 1155 (1985) and references therein
       A.G. Cullis, H.C. Webber, J.M. Poate, A.L. Simons: Appl. Phys. Lett. **36**, 320 (1980)
2.72   L. Eckertova: *Physics of Thin Films* (Plenum, London 1984)
2.73   J.A. Venables: Philos. Mag. **27**, 697 (1973)
2.74   K. Takeuchi, K. Kinosita: Thin Solid Films **90**, 31 (1982)
2.75   G. Zinsmeister: Vacuum **16**, 529 (1966)
2.76   D.J. Ehrlich, J.Y. Tsao: SPIE **4959**, 2 (1984)
2.77   R.B. Gold, J.F. Gibbons: J. Appl. Phys. **51**, 1256 (1980)
2.78   Z.L. Liau: Appl. Phys. Lett. **34**, 221 (1979)
2.79   R.B. Gold, J.F. Gibbons: In *Laser and Electron Beam Processing of Materials*, ed. by C.W. White, P.S. Peercy (Academic, New York 1980) p77
2.80   A. Lietoila, R. Gold, J.F. Gibbons: Appl. Phys. Lett. **39**, 810 (1981)
2.81   H.H. Gilgen, C.J. Chen, R. Krchnavek, R.M. Osgood Jr: In *Laser Processing and Diagnostics*, ed. by D. Bäuerle, Springer Ser. Chem. Phys., Vol.39 (Springer, Berlin, Heidelberg 1984) p.225
2.82   W.M. Donnelly, M. Greva, J. Long, R.F. Karlicek: Appl. Phys. Lett. **43**, 454 (1983)
2.83   J. Zavelovich, M. Rothschild, W. Gornik, C.K. Rhodes: J. Chem. Phys. Soc. **74**, 6787 (1981)
2.84   M. Zelikoff, K. Watanabe, E.C.Y. Inn: J. Chem. Phys. **21**, 1643 (1953)
2.85   *American Institute of Physics Handbook*, 2nd ed., ed. by D.E. Gray (AIP, New York 1972)
2.86   *Handbook of Optics*, ed. by W.D. Driscol, W. Vaughan (McGraw-Hill, New York 1978)
2.87   *Pulsed Laser Processing of Semiconductors*, ed. by R.F. Wood, C.W. White, R.T. Young (Academic, New York 1984)

2.88 A.L. Smirl, T.F. Boggess, S.C. Moss, I.W. Boyd: J. Lumines-
cence **30**, 272 (1985)
2.89 W.C. Dash, R. Newman: Phys. Rev. **99**, 1151 (1955)
2.90 H.Y. Fan, W.G. Spitzer, R.J. Collins: Phys. Rev. **101**, 566
(1956)
2.91 J.M. Poate, J.W. Mayer (eds.): *Laser Annealing of Semiconductors* (Academic, New York 1982)
2.92 P.W. Baumeister: Phys. Rev. **121**, 359 (1961)
2.93 J.E. Eby, K.J. Teegarden, D.B. Dutton: Phys. Rev. **116**, 1099
(1959)
2.94 I.W. Boyd, J.I.B. Wilson: J. Appl. Phys. **53**, 4166 (1982)
2.95 I.W. Boyd: In *Surface Studies with Lasers*, ed. by F.R. Aussenegg, A. Leitner, M.E. Lippitsch, Springer Ser. Chem.
Phys., Vol.33 (Springer, Berlin, Heidelberg 1983) p.193
2.96 I.W. Boyd: In *Dielectric Layers in Semiconductors*, ed. by
G.G. Bentini, E. Fogarassy, A. Golanski (Les Editions de
Physique, Les Ulis 1986) p.177
2.97 R.J. Elliott, A.F. Gibson: *An Introduction to Solid State
Physics and its Applications* (MacMillan, London 1982)
2.98 C. Kittel: *Introduction to Solid State Physics* (Wiley,
London 1972)
2.99 R.A. Smith: *Semiconductors* (Cambridge, London 1979)
2.100 C.N. Banwell: *Fundamentals of Molecular Spectroscopy*
(McGraw-Hill, London 1972)
2.101 B.G. Streetman: *Solid State Electronic Devices* (Prentice
Hall, London, 1980)
2.102 J.I. Pankove: *Optical Processes in Semiconductors* (Dover,
New York 1971)
2.103 S.M. Sze: *Physics of Semiconductor Devices* (Wiley, New
York 1981)
2.104 M.C. Downer, R.L. Fork, C.V. Shank: J. Opt. Soc. Am. B **2**,
595 (1985)
2.105 J.M. Liu, R. Yen, H. Kurz, N. Bloembergen: Appl. Phys.
Lett. **39**, 755 (1981)
2.106 P.H. Bucksbaum, J. Bokor: In *Energy Beam–Solid Interactions and Transient Thermal Processing*,ed. by J.C.C. Fan,
N.H. Johnson (North–Holland, Amsterdam 1984) p.93

Recommendations for Further Reading

Allmen, M.von: *Laser–Beam Interactions with Materials*, Springer
Ser. Mat. Sci., Vol.2 (Springer, Berlin, Heidelberg 1987)

D. Bäuerle: *Chemical Processing with Lasers*, Springer Ser. Mat.
Sci., Vol.1 (Springer, Berlin, Heidelberg 1986)

Madelung, O.: *Introduction to Solid-State Theory*, Springer Ser.
Solid-State Sci., Vol.2 (Springer, Berlin, Heidelberg 1978)

Seeger K.: *Semiconductor Physics*, 3rd. ed., Springer Ser. Solid-
State Sci., Vol.40 (Springer, Berlin, Heidelberg 1985)

Chapter 3

3.1   H. Kogelnik, T. Li: Appl. Opt. **5**, 1550 (1966)
      H.K.V. Lotsch: Optik **30**, 1, 181, 217, 563 (1969/70)
3.2   A. Yariv: *Quantum Electronics*, (Wiley, New York 1967)
3.3   D.J. Ehrlich, J.Y. Tsao: J. Vac. Sci. Technol. B1, 969
      (1983)
3.4   D.V. Podlesnik, H.H. Gilgen, R.M. Osgood: In *Laser Con-
      trolled Chemical Processing of Surfaces*, ed. by A.W. John-
      son, D.J. Ehrlich (Elsevier, Amsterdam 1984) p.161
3.5   D.J. Ehrlich, R.M. Osgood, T.F. Deutsch: Appl. Phys. Lett.
      39, 957 (1981)
3.6   B.M. McWilliams, I.P. Herman, F. Mitlitsky, R.A. Hyde, L.L.
      Wood: Appl. Phys. Lett. **43**, 946 (1983)
3.7   D.J. Ehrlich, R.M. Osgood Jr, T.F. Deutsch: IEEE J. QE-16,
      1233 (1980)
3.8   R. Solomon, L.F. Mueller: US Patent 3,364,087 (January
      1968)
3.9   M. Murahara, K. Toyoda: In *Laser Processing and Diagnos-
      tics*, ed. by D. Bäuerle, Springer Ser. Chem. Phys., Vol. 39
      (Springer, Berlin, Heidelberg 1984) p.252
3.10 I.W. Boyd: In *Laser Processing and Diagnostics*, ed. by D.
      Bäuerle, Springer Ser. Chem. Phys., Vol. 39 (Springer,
      Berlin, Heidelberg 1984) p.274
3.11 H. Sankur, J.T. Cheung: J. Vac. Sci. Technol. A1, 1806
      (1983)
3.12 K. Jain, C.G. Wilson, B.J. Lin: IBM J. Res. Dev. **26**, 151
      (1982)
3.13 K. Jain: Lasers & Appl. **2**, 49 (September 1983)
3.14 D. Henderson, J.C. White, H.G. Craighead, I. Adesida: Appl.
      Phys. Lett. **46**, 900 (1985)
3.15 H.G. Craighead, J.C. White, R.E. Howard, L.D. Jackel, R.E.
      Behringer, J.E. Sweeney, R.W. Epworth: J. Vac. Sci. Tech-
      nol. B1, 1186 (1983)
3.16 J.C. White, H.G. Craighead, R.E. Howard, L.D. Jackel, R.E.
      Behringer, R.W. Hepworth, D. Henderson, J.E. Sweeney:
      Appl. Phys. Lett. **44**, 22 (1984)
3.17 G.L. Loper, M. Tabat: SPIE **459**, 121 (1984)
3.18 S.M. Metev, S.K. Savtchenko, K. Stamenov: J. Phys. D13,
      L75 (1980)
3.19 M. Latta, R. Moore, S. Rice, K. Jain: J. Appl. Phys. **56**,
      587 (1984)
3.20 D.J. Ehrlich, J.Y. Tsao, C.O. Bozler: J. Vac. Sci. Technol.
      B3, 1 (1985)
3.21 I.W. Boyd, T.F. Boggess, S.C. Moss, A.L. Smirl: In *Laser
      Processing and Diagnostics*, ed. by D. Bäuerle, Springer
      Ser. Chem. Phys., Vol.39 (Springer, Berlin, Heidelberg
      1984) p.50
3.22 A.G. Cullis, H.C. Weber, N.G. Chew: J. Phys. E12, 688
      (1979)
3.23 R.E. Grojean, D. Feldman, J.F. Roach: Rev. Sci. Instrum.
      51, 375 (1980)

3.24 M. Lacombat, G.M. Dubroeucq, J. Massin, M. Brevignon: Solid State Technology p.115 (August 1980)
3.25 Y. Kawamura, Y. Itagaki, K. Toyoda, S. Namba: Opt. Commun. **48**, 44 (1983)
3.26 A. Brunsting: Redirecting Surface for Desired Intensity Profile, U.S.Patent 4,327,972 (May 1982)
3.27 Spawr Optical Research, 1527 Pomona Road. Corona, CA 91720, USA
3.28 M.R. Latta, K. Jain: Opt. Commun. **49**, 435 (1984)
3.29 M.J. Soileau: IEEE J. QE-20, 464 (1984) and references therein.
3.30 F. Keilman, Y.H. Bai: Appl. Phys. A29, 9 (1982)
3.31 S.R.J. Brueck, D.J. Ehrlich: Phys. Rev. Lett. **48**, 1678 (1982)
3.32 Z. Guosheng, P.M. Fauchet, A.E. Siegman: Phys. Rev. B26, 5366 (1982)
3.33 R.J. Wilson, F.A. Houle: Phys. Rev. Lett. 55, 2184 (1985)
3.34 A.L. Smirl, I.W. Boyd, T.F. Boggess, S.C. Moss, H.M. van Driel: J. Appl. Phys. **60**, 1169 (1986) and references therein
3.35 H.M. van Driel, J.E. Sipe, J.F. Young: Phys. Rev. Lett. **49**, 1955 (1982)
3.36 J.F. Young, J.E. Sipe, H.M. van Driel: Phys. Rev. B30, 2001 (1984)
3.37 D. Jost, W. Luthy, H.P. Weber: Proc. E-MRS, Symp.B (Les Editions de Physique, Les Ulis 1986)
3.38 H.J. Kreuzer: U.S. Patent 3,476,463 (1969)
3.39 D. Shafer: Optics and Laser Technology **14**, 159 (1982)
3.40 I.W. Boyd, J.G. Crowder: J. Phys. E15, 421 (1982)
3.41 I.W. Boyd: Optics and Laser Technology 15, 150 (1983)
3.42 G.C. Lim, W.M. Steen: Optics and Laser Technology **14**, 149 (1982)
3.43 C.K. Rhodes (ed.): *Excimer Lasers*, 2nd ed., Topics Appl. Phys., Vol.30 (Springer, Berlin, Heidelberg 1984)
3.44 R. Sauerbrey, H. Langhoff: IEEE J. QE-21, 179 (1985)

Recommendations for Further Reading

Barbe D.F. (ed.): *Very Large Scale Integration VLSI*, 2nd. ed., Springer Ser. Electrophys., Vol 5 (Springer, Berlin, Heidelberg 1982)

Einspruch N.G. (ed.): Principles of Optical Lithography, in *VLSI Electronics Microstructure Science*, Vol.1 (Academic, New York 1981)

Marcuse D.: *Light Transmission Optics* (Van Nostrand, Princeton 1973)

Newman R. (ed.): *Fine Line Lithography* (North-Holland, Amsterdam 1980)

Siegman A.E.: *Introduction to Lasers and Masers* (McGraw–Hill, New York 1968)

Yariv A.: *Introduction to Optical Electronics* (Holt, Rinehart and Winston, New York 1971)

Chapter 4

4.1 N. Cabrera, N.F. Mott: Rep. Prog. Phys. **12**, 163 (1948/1949)
4.2 F.P. Fehlner, N.F. Mott: Oxidation in Metals 2, 59 (1970)
4.3 E.J.W. Verwey: Physica 2, 1059 (1932)
4.4 C. Wagner: Z. Physik. Chem. B **32**, 447 (1936)
4.5 B.E. Deal, A.S. Grove: J. Appl. Phys. **36**, 3770 (1965)
4.6 I.W. Boyd, J.I.B. Wilson: J. Appl. Phys. **53**, 4166 (1982)
4.7 H. Ibach, K. Horn, R. Dorn, H. Lüth: Surf. Sci. **38**, 433 (1973)
4.8 C.M. Garner, I. Lindau, C.Y. Su, P. Pianetta, W. Spicer: J. Vac. Sci. Technol. **4**, 372 (1977)
4.9 R. Karchner, L. Ley: Solid State Commun. **43**, 415 (1982)
4.10 C. Camelin, G. Demazeau, A. Straboni, J.L. Buevoz: Appl. Phys. Lett. **48**, 1211 (1986)
4.11 D.N. Modlin, W.A. Tiller: J. Electrochem. Soc. **132**, 1659 (1985)
4.12 L.M. Landsberger, W.A. Tiller: Appl. Phys. Lett. **49**, 143 (1986)
4.13 M.A. Hopper, R.A. Clark, L. Young: J. Electrochem. Soc. **122**, 1216 (1975)
4.14 E.A. Irene, Y.J. van der Meulen: J. Electrochem. Soc. **123**, 1380 (1976)
4.15 Y. Kamigaki, Y. Itoh: J. Appl. Phys. **48**, 2891 (1977)
4.16 A.C. Adams, T.E. Smith, C.C. Chang: J. Electrochem. Soc. **127**, 1787 (1980)
4.17 A.G. Revesz, R.J. Evans, J. Phys. Chem. Sol. **30**, 551 (1969)
4.18 E.A. Irene: J. Appl. Phys. **54**, 5416 (1983)
4.19 R.H. Doremus: Thin Solid Films **122**, 191 (1984)
4.20 A. Fargeix, G. Ghibaudo, G. Kamarinos: J. Appl. Phys. **54**, 2878 (1983)
4.21 A.S. Grove: *Physics and Technology of Semiconductor Devices* (Wiley, New York 1967)
4.22 W.A. Tiller: J. Electrochem. Soc. **128**, 689 (1981)
4.23 A. Lora–Tamayo, E. Dominguez, E. Lora–Tamayo, J. Llabres: Appl. Phys. **17**, 79 (1978)
4.24 S.M. Hu: Appl. Phys. Lett. **42**, 872 (1983)
4.25 Y.Z. Lu, Y.C. Cheng: J. Appl. Phys. **56**, 1608 (1984)
4.26 S.A. Schafer, S.A. Lyon: Appl. Phys. Lett. **47**, 154 (1985)
4.27 R. Ghez, Y.J. van der Meulen: J. Electrochem. Soc. **119**, 1100
4.28 J. Blanc: Appl. Phys. Lett. 33, 424 (1978)
4.29 W.A. Tiller; J. Electrochem. Soc. **119**, 591 (1980)
4.30 P.J. Jorgensen: J. Chem. Phys. **37**, 874 (1962)
4.31 D.O. Raleigh: J. Electrochem. Soc. **113**, 782 (1966)

4.32 W.A. Tiller: J. Electrochem. Soc. **127**, 619 (1980)
4.33 T.G. Mills, F.A. Kroger: J. Electrochem. Soc. **12**, 1582 (1973)
4.34 N.F. Mott, Phil.Mag. **55**, 117 (1987)
4.35 M.M. Atalla: In *Properties of Elemental and Compound Semiconductors*, ed. by H. Gatos (Interscience, New York 1960) Vol 5, pp.163–181
4.36 J.R. Ligenza, W.G. Spicer: J. Phys. Chem. Solids **14**, 131 (1960)
4.37 E. Rosencher, A. Straboni, S. Rigo, G. Amsel: Appl. Phys. Lett. **34**, 254 (1979)
4.38 F. Rochet, B. Agius, S. Rigo: J. Electrochem. Soc. **121**, 914 (1984)
4.39 J.C. Mikkelsen: Appl. Phys. Lett. **45**, 1189 (1985)
4.40 N.F. Mott: Phil. Mag. A **45**, 323 (1982)
4.41 W.A. Tiller: J. Electrochem Soc. **127**, 625 (1980)
4.42 J.F. Asmus, F.S. Baker: Record of 10th Symp. on Electron, Ion, and Laser Beam Technology San Francisco Press, San Francisco (1969) p.241
4.43 A.G. Akimov, A.P. Gagarin, V.G. Dugarov, V.S. Makin, S.D. Pudkov: Sov. Phys. Tech. Phys. **25**, 1439 (1980)
4.44 M.I. Arzuov, A.I. Barchukov, F.V. Bunkin, N.A. Kirichenko, V.I.B.S. Luk'yanchuk: Sov. J. Quantum Electron. **9**, 281 (1979)
4.45 V.P. Veiko, G.A. Kotov, M.N. Libenson, M.N. Nikitin: Sov. Phys. Doklady **18**, 83 (1973)
4.46 M.I. Arzuov, A.I. Barchukov, F.V. Bunkin, V.I. Kónov, A.A. Lyubin: Sov. J. Quant. Electron. **5**, 931 (1976)
4.47 I.W. Boyd: Contemp. Phys. **24**, 461 (1983)
4.48 I. Ursu, L.Nanu, M. Dinescu, Al. Hening, I.N. Mihailescu, L.C. Nistor, V.S. Teodorescu, E. Szil, I. Hevesi, J. Kovacs, L. Nanai: Appl. Phys. A **35**, 103, (1984)
4.49 I. Ursu, L.C. Nistor, V.S. Teodorescu, I.N. Mihailescu, L. Nanu, A.M. Prokhorov, N.I. Chaplieu, V.I. Konov: Appl. Phys. Lett. **44**, 188 (1984)
4.50 I. Ursu, L. Nanu, I.N. Mihailescu, L.C. Nistor, V.S. Teodorescu, A.M. Prokhorov, V.I. Konov, N.I. Chaplieu: J. Physique. Lett. **45**, L737 (1984)
4.51 M. Wautelet, L. Baufay: Thin Solid Films **100**, L9 (1983)
4.52 M. Wautelet: Mat. Lett. **2**, 20 (1984)
4.53 R. Andrew: In *Interfaces Under Laser Radiation*, ed. by L.D. Laude, D. Bäuerle, M. Wautelet (Nijhoff, The Hague 1987)
4.54 R. Merlin, T.A. Perry: Appl. Phys. Lett. **45**, 852 (1984)
4.55 A.G. Akimov, A.M. Bonch-Bruevich, A.P. Gargarin, V.G. Dorofeev, P.A. Zimin, I.N. Ivanova, M.N. Libenson, V.S. Makin, S.D. Pudkov: Bull. Acad. USSR Phys. Ser. **46**, 145 (1982)
4.56 M. Thuillard, M. von Allmen: Appl. Phys. Lett. **47**, 936 (1985)
4.57 R.F. Marks, R.A. Pollak: J. Chem. Phys. **81**, 1019 (1984)
4.58 R.F. Marks, R.F. Pollack, Ph. Avouris,C.T. Lin, Y.J. Thefaine: J. Chem. Phys. **78**, 4270 (1983) and references therein

4.59   S.M. Metev, S.K. Savtchenko, K. Stamenov: J. Phys. D **13**, L75 (1980)

4.60   S.T. Pantelides: *The Physics of SiO$_2$ and Its Interfaces* (Pergamon, New York 1978) and references therein

4.61   B.E. Deal: In *Semiconductor Silicon*, ed. by H.R. Huff, E.Sirtl (Electrochem. Soc., Princeton 1977) p.276

4.62   L.I. Maissel: In *Handbook of Thin Film Technology*, ed. by L.I. Maissel, R. Glang (McGraw-Hill, New York 1970) Chap.4

4.63   J.L. Vossen: J. Vac. Sci. Technol. **8**, 512 (1971)

4.64   W.A. Pliskin, R.A. Gdula: In *Handbook on Semiconductors,* ed. by T.S. Moss (North-Holland, Amsterdam 1982) p.641

4.65   K. Urbanek: Solid State Technol. **20**, 87 (1977)

4.66   See, e.g., S. Dzioba, G. Este, H.M. Naguib: J. Electrochem. Soc. **129**, 2537 (1982) and references therein

4.67   V.Q. Ho: Jpn. J. Appl. Phys. Suppl.1, **19**, 103 (1983)

4.68   T. Tokuyama, S. Kimura, T. Warabisako, E. Murakami, K. Miyake: In *Laser Processing and Diagnostics*, ed. by D. Bäuerle, Springer Ser. Chem. Phys., Vol.39 (Springer, Berlin, Heidelberg 1984) p.288

4.69   J.R. Ligenza, W.G. Spicer: J. Phys. Chem. Solids **14**, 132 (1960)

4.70   R.J. Zeto, N.O. Korolkoff, S. Marshall: Solid State Techn. **22**, 62 (July 1979)

4.71   J.F. Gibbons: Jpn. J. Appl. Phys., Suppl., **19**, 121 (1981)

4.72   I.W. Boyd, J.I.B. Wilson, J.L. West: Thin Solid Films **83**, L173 (1981)

4.73   S.A. Schafer, S.A. Lyon: J. Vac. Sci. Technol. **19**, 494 (1981)

4.74   I.W. Boyd, J.I.B. Wilson: Appl. Phys. Lett. **41**, 162 (1982)

4.75   S.A. Schafer, S.A. Lyon: J. Vac. Sci. Technol. **21**, 422 (1982)

4.76   E.M. Young, W.A. Tiller: Appl. Phys. Lett. **42**, 63 (1983)

4.77   I.W. Boyd: Appl. Phys. Lett. **42**, 728 (1983)

4.78   I.W. Boyd, T.D. Binnie, J.I.B. Wilson, M.J. Colles: J. Appl. Phys. **55**, 3061 (1984)

4.79   H.S. Carslaw, J.C. Jaeger: *Conduction of Heat in Solids* (Clarendon, Oxford 1959)

4.80   H.W. Ho, A. Compaan: J. Appl. Phys. **51**, 1565 (1980)

4.81   I.W. Boyd: In *Dielectric Layers in Semiconductors*, ed. by G.G. Bentini, E. Fogarassy, A. Golanski (Les Editions de Physique, Les Ulis 1986) p.177

4.82   I.W. Boyd, S.C. Moss: In Laser Chemical Processing of Semiconductor Devices, Extended abstracts of MRS Symp.B, ed. by F.A. Houle, T.F. Deutsch, R.M. Osgood (Materials Research Society, Pittsburgh 1984) p.132

4.83   I.W. Boyd, F. Micheli: Electron. Lett. **23**, 298 (1987)

4.84   P.A. Bertrand: J. Vac. Sci. Technol. **132**, 973 (1985)

4.85   J. Siejka, J. Perriere, R. Srinivasan: Appl. Phys. Lett. **46**, 773 (1985)

4.86   K.A. Bertness, W.G. Petro, J.A. Silberman, D.J. Friedman, W.E. Spicer: J. Vac. Sci. Technol. A 3, 1464 (1985)

4.87 W.G. Petro, I. Hino, S. Eglash, I. Lindau, C.Y. Su, W.E. Spicer: J. Vac. Sci. Technol. **21**, 405 (1982)

4.88 V.M. Bermudez: J. Appl. Phys. **54**, 6795 (1983)

4.89 M. Fukuda, K. Takahei: J. Appl. Phys. **57**, 129 (1985)

4.90 M. Fathipour, P.K. Boyer, G.J. Collins, C.W. Wilmsen: J. Appl. Phys. **57**, 637 (1985)

4.91 S.E. Blum, K. Brown, R. Srinivasan: Appl. Phys. Lett. **43**, 1026 (1983)

4.92 T. Sugii, T. Ho, H. Ishikawa Appl. Phys. Lett. **45**, 966 (1984)

4.93 T.J. Chuang, I. Hussla, W. Sesselmann: In *Laser Processing and Diagnostics*, ed. by D. Bäuerle, Springer Ser. Chem. Phys., Vol.39 (Springer, Berlin, Heidelberg 1984) p.300

4.94 C.I.H. Ashby: Appl. Phys. Lett. **43**, 609 (1983)

4.95 I.W. Boyd: In *Interfaces Under Laser Radiation*, ed. by L.D. Laude, D. Bäuerle, M. Wautelet (Martinus Nijhoff, The Hague 1987) p.409

4.96 R. Oren, S.K. Ghandi: Appl. Phys. Lett. **42**, 752 (1971)

4.97 C. Fiori: Phys. Rev. Lett. **52**, 2077 (1984)

4.98 R.T. Young, J. Narayan, W.H. Christie, G.A. van der Leeden, D.E. Rothe, L.J. Chen: Solid State Techn. **26**, 183 (1983)

4.99 I.W. Boyd: J. Appl. Phys. **54**, 3561 (1983)

4.100 W.A. Pliskin, H.S.Lehmann: J. Electrochem. Soc. **112**, 1013 (1965)

4.101 M. Nakamura, Y. Mochizuki, K. Usami, Y. Itoh, T. Nozaki: Solid State Commun. **50**, 1079 (1984)

4.102 E. Hensel, K. Wollschläger, D. Schulze, U. Kreissig, W. Skorupa, J. Finster: Surface and Interface Analysis **7**, 205 (1985)

4.103 I.W. Boyd, J.I.B. Wilson: Appl. Phys. Lett. **50**, 320 (1987)

4.104 T.E. Orlowski, H. Richter: Appl. Phys. Lett. **45**, 241 (1984)

4.105 H. Zarnani, P.K. Boyer, M. Fathipour, C.W. Wilmsen, R. Solanki, G.J. Collins: In *Laser Chemical Processing of Semiconductor Devices*, ed. by F.A. Houle, T.F. Deutsch, R.M. Osgood (MRS, Pittsburgh 1984) p.122

4.106 C.M. Gronet, J.C. Sturm, K.E. Williams, J.F. Gibbons: In *Rapid Thermal Processing*, ed. by T.O. Sedgwick, T.E. Seidel, B.-Y. Tsaur (Materials Research Society, Pittsburgh 1986) p.305

4.107 J.C. Gelpey, P.O. Stump, R.A. Capodilupo: In *Rapid Thermal Processing*, ed. by T.O. Sedgwick, T.E. Seidel, B.-Y. Tsaur (Materials Research Society, Pittsburgh 1986) p.321

4.108 Z.A. Weinberg, T.N. Nguyen, S.A. Cohen, R. Kalish: In *Rapid Thermal Processing*, ed. by T.O. Sedgwick, T.E. Seidel, B.-Y. Tsaur (Materials Research Society, Pittsburgh 1986) p.327

4.109 J. Nulman, J.P. Krusius, N. Shah, A. Gat, A. Baldwin: J. Vac. Sci. Technol. A **4**, 1005 (1986)

4.110 J.P. Ponpon, J.J. Grob, A. Grob, R. Stuck: J. Appl. Phys. **59**, 3921 (1986)

4.111 N. Chan Tung, Y. Caratini, L. Liauzu: In *Dielectric Layers in Semiconductors*, ed. by G.G. Bentini, E. Fogarassy, A.

Golanski (Les Editions de Physique, Les Ulis 1986) p.247
4.112 J. Nulman, J.P. Krusius, A. Gat: IEEE EDL-6, 205 (1985)
4.113 M.M. Moslehi, S.C. Shatas, K.C. Saraswat: Appl. Phys. Lett. **47**, 1353 (1985)
4.114 M.M. Moslehi, K.C. Saraswat: IEEE J. SC-20, 26 (1985)
4.115 M.M. Moslehi, K.C. Saraswat, S.C. Shatas: Appl. Phys. Lett. **47**, 1113 (1985)
4.116 H.Z. Massoud, J.D. Plummer, E.A. Irene: J. Electrochem. Soc.**132**, 1745 (1985)
4.117 A.M. Hodge, C. Pickering, A.J. Pidduck, R.W. Hardeman: In *Rapid Thermal Processing*, ed. by T.O. Sedgwick, T.E. Seidel, B.-Y. Tsaur (Materials Research Society, Pittsburgh 1986) p.313
4.118 J.J. Bentini, M. Berti, C.Cohen, A.V. Drigo, E. Jannitti, D. Pribat, J. Siejka: J. Physique **43**, C1-229 (1982)
4.119 C. Cohen, J. Siejka, M. Berti, A.V. Drigo, G.G. Bentini, D. Pribat, E. Jannitti: J. Appl. Phys. **55**, 4081 (1984)
4.120 K.A. Bertness, W.G. Petro, J.A. Silberman, D.J. Friedman, T. Kendelwicz, W.E. Spicer: In *Laser Chemical Processing of Semiconductor Devices*, ed. by F.A. Houle, T.F. Deutsch, R.M. Osgood (MRS, Pittsburgh 1984) p.129
4.121 T. Ito, T. Nozaki, H. Arkawa, M. Shinoda: Appl. Phys. Lett. **32**, 330 (1978)
4.122 R.P.H. Chang, C.C. Chang, S. Darak: Appl. Phys. Lett. **36**, 999 (1980)
4.123 T. Ito, I. Kato, T. Nozaki, T. Nakamura, H. Ishikawa: Appl. Phys. Lett. **38**, 370 (1981)
4.124 H. Nakamura, M. Kaneko, S. Matsumoto, S. Fukita, A. Sasaki: Appl. Phys. Lett. **43**, 691 (1983)
4.125 S. Wong, W. Oldham: IEEE EDL-5, 175 (1984)
4.126 M. Hirayama, T. Matsukawa, H. Arima, Y. Ohno, N. Tsubouchi, H. Nakata: J. Electrochem. Soc. **131**, 663 (1984)
4.127 C.Y. Wu, D.W. King, M.K. Lee, C.T. Chen: J. Electrochem. Soc. **129**, 1559 (1982)
4.128 I. Ursu, I. N. Mihailescu, L. Nanu, A.M. Prokhorov, V.I. Konov, V.G. Ralchenko, Appl. Phys. Lett. **46**, 110 (1985)
4.129 I. Ursu, I.N. Mihailescu, A.M. Prokhorov, V.I. Konov, V.N. Tokarev, S.A. Uglov: J. Phys. D **18**, 2547 (1985)
4.130 I. Ursu, I.N. Mihailescu, A.M. Prokhorov, V.I. Konov: *Trends in Quantum Electronics*, ed. by I.Ursu, A.M. Prokhorov (Springer, Berlin, Heidelberg; Institute of Physics, Bucharest 1986) and references therein.
4.131 R.R. Krchnavek, H.H. Gilgen, R.M. Osgood: J. Vac. Sci. Technol. B 2, 641 (1984)
4.132 T. Nakayama, M. Okigawa, N. Itoh: Nucl. Inst. Meth. Phys. Res. B 1, 301 (1984)
4.133 M. Matsuura, M. Ishida, A. Suzuki, K. Hara: Jpn. J. Appl. Phys. **20**, L726 (1981)
4.134 J.T. Yardley, A. Gupta, G. West, K.W. Beeson: In *Laser Chemical Processing of Semiconductor Devices*, ed. by F.A.

Houle, T.F. Deutsch, R.M. Osgood, Extended Abstracts of Symp. B (MRS, Pittsburgh 1984) p.121

4.135 R. Solanki, W. Ritchie, G.J. Collins: Appl. Phys. Lett. **43**, 454 (1983)

4.136 T.F. Deutsch, D.J. Silversmith, R.W. Mountain: In *Laser Diagnostics and Photochemical,Processing for Semiconductor Devices*, ed. by R.M. Osgood, S.R.J. Brueck, H.R. Schlossberg (North-Holland, Amsterdam, 1983) p.129

4.137 W.M. Donnelly, M. Geva, J. Long, R.F. Karlicek: Appl. Phys. Lett. **44**, 951 (1984)

4.138 S. Nishino, H. Honda, H. Matsunami: Jpn. J. Appl. Phys.**25**, L87 (1986)

4.139 S.D. Allen: SPIE **198**, 49 (1979)

4.140 R. Solanki, G.J. Collins: Appl. Phys. Lett. **42**, 662 (1983)

4.141 H. Sankur, J.T. Cheung: J. Vac. Sci. Technol. A 1, 1806 (1983)

4.142 H. Sankur: SPIE **459**, 78 (1984)

4.143 G.G. Bentini, M. Berti, A.V. Drigo, E. Jannitti, C. Cohen, J. Siejka: In *Laser Chemical Processing of Semiconductor Devices*, ed. by F.A. Houle, T.F. Deutsch, R.M. Osgood (MRS, Pittsburgh 1984) p.126

4.144 P.K. Boyer, G.A. Roche, W.H. Ritchie, G.J. Collins: Appl. Phys. Lett. **40**, 716 (1982)

4.145 P.K. Boyer, K.A. Emery, H. Zarnani, G. J. Collins: Appl. Phys. Lett. **45**, 979 (1984)

4.146 S. Szikora, W. Krauter, D. Bäuerle: Mat. Lett. **2**, 263 (1984)

4.147 Y. Hirota, O. Mikami: Electron. Lett. **21**, 77 (1985)

4.148 S.W. Chiang, Y.S. Liu, R.F. Reihl: In *Laser and Electron Beam Solid Interactions and Materials Processing*, ed. by J.F. Gibbons, L.D. Hess, T.W. Sigmon (Elsevier, New York 1981)

4.149 A. Garulli, M. Servidori, I. Vecchi: J. Appl. Phys. D **13**, L199 (1981)

4.150 E. Fogarassy, C.W. White, D.H. Lowndes, J. Narayan: In *Beam-Solid Interactions and Phase Transformations*, ed. by H. Kurz, G.L. Olson, J.M. Poate (MRS, Pittsburgh 1986) p.173

4.151 I.W. Boyd: Appl. Phys. A **31**, 71 (1983)

4.152 K. Hoh, H. Koyama, K. Uda, Y. Miura: Jpn. J. Appl. Phys. **19**, L375 (1980)

4.153 K. Tamagawa, T. Hayashi, S. Komiya: Jpn. J. Appl. Phys. **25**, L728 (1986)

4.154 A. Sugimura, M. Hanabusa: Jpn. J. Appl. Phys. **26**, L56 (1987)

4.155 A. Cros, R. Salvan, J. Derrien: Appl. Phys. A **28**, 241 (1982)

4.156 Y.S. Liu, S.W. Chiang, F. Bacon: Appl. Phys. Lett. **38**, 1005 (1981)

4.157 Y.S. Liu, S.W. Chiang, W. Katz: In *Laser and Electron Beam Interactions with Solids*, ed. by B.R. Appleton, G.K. Celler (Elsevier, New York 1982)

300

4.158 M. Morita, S. Aritomi, T. Tanaka, M. Hirose: Appl. Phys.
Lett. **49**, 699 (1986)
4.159 M.I. Birjega, L. Nanu, I.N. Mihailescu, M. Dinescu, N. Pop-
escu-pogrion, C. Sarbu: Optica Acta **33**, 1073 (1986)
4.160 E. Fogarassy, S. Unamuno, J.L. Regolini, C. Fuchs: Phil.
Mag. B **55**, 253 (1987)
4.161 R. Andrew: Appl. Phys. B **41**, 205 (1986)
4.162 I. Ursu, L. Nanu, I.N. Mihailescu: Appl. Phys. Lett. **49**,
109 (1986)
4.163 J.M. Jasinski, B.S. Meyerson, T.N. Nguyen: J. Appl. Phys.
61, 431 (1987)
4.164 S.B. Ogale, P.P. Patil, S. Roorda, F.W. Saris: Appl. Phys.
Lett. **50**, 1802 (1987)
4.165 E.T.S. Pan, J.H. Flint, D. Adler, J.S. Haggerty: J. Appl.
Phys. **61**, 4535 (1987)
4.166 R.A.P. Devine, G. Auvert: Appl. Phys. Lett. **49**, 1605 (1986)

Recommendations for Further Reading

Kelly M.J., C. Weisbuch (eds.): *The Physics and Fabrication of
Microstructures and Microdevices*, Springer Proc. Phys., Vol.13
(Springer, Berlin, Heidelberg 1986)

Le Lay G., J. Derrien, N. Boccara (eds.): *Semiconductor Interfaces: Formation and Properties*, Springer Proc. Phys., Vol.22
(Springer, Berlin, Heidelberg 1987)

Nicollian E.H., J.R. Brews: *MOS Physics and Technology* (Wiley,
New York, 1982)

Nizzoli F., K.-H. Rieder, R.F. Willis (eds.): *Dynamical Phenomena
at Surfaces, Interfaces and Superlattices*, Springer Ser. Surf.
Sci., Vol.3 (Springer, Berlin, Heidelberg 1985)

Sedgewick T.O., T.E. Seidel, B.-Y. Tsaur (eds.): *Rapid Thermal
Processing*, Proc. MRS **52** (MRS, Pittsburgh 1986)

Chapter 5

5.1 J.M. Fairfield, G.H. Schwuttke: Solid State Elect. **11**,
1175 (1968)
5.2 M. Birnbaum, T.L. Stocker: J. Appl. Phys. **39**, 6032 (1968)
5.3 J. Feinlab, J. de Neufville, S.C. Moss, S.R. Ovshinski: J.
Appl. Phys. **18**, 254 (1971)
5.4 R.J. von Gutfield, P. Chandhari: J. Appl. Phys. **43**, 4688
(1972)
5.5 Yu. A. Bykovskii, A.G. Dudoladov, V.P. Kozlenkov, P.A.
Leont'ev: JEPT Lett. **20**, 135 (1974)
5.6 A.A. Agasiev, A.K.H. Zeinally, V.M. Salmanov, V.I. Tagirov,
K.Yu Karakurchi: Sov. Phys. Semicond. **9**, 777 (1975)

5.7   I.B. Khaibullin, E.I. Shtyrkov, M.M. Zaripov, M.F. Galyaut-dinov, R.M. Bajazitov: Paper No. 2061-74, deposited in Viniti, Moscow (1974) (in Russian)

5.8   I.B. Khaibullin, E.I. Shtyrkov, M.M. Zaripov, R.M. Bay-zitov, M.F. Galjanudinov: Rad. Eff. **36**, 225 (1978) and references therein

5.9   *Energy Beam Solid Interactions and Transient Processing*, ed. by D.K. Biegelsen, G. Rozgonyi, C.V. Shank (Materials Research Society, Pittsburgh 1984) and references therein

5.10 J.M. Poate, J.W. Mayer (eds.): *Laser Annealing of Semiconductors* (Academic, New York 1982)

5.11 I.W. Boyd, J.I.B. Wilson: Nature **303**, 481 (1983)

5.12 I.W. Boyd: Contemp. Phys. **24**, 461 (1983)

5.13 A.G. Cullis: Rep. Prog. Phys. **48**, 1156 (1986)

5.14 G. Battaglin, G. Della Mea, A.V. Drigo, G. Foti, G.G. Bentini, M. Servidori: Phys. Stat. Sol. **49**, 347 (1978)

5.15 P.L. Liu, R. Yen, N. Bloembergen, R.T. Hodgson: Appl. Phys. Lett. **34**, 864 (1979)

5.16 R. Tsu, R. Hodgson, T.Y. Tan, J.E. Baglin: Phys. Rev. Lett. **42**, 1356 (1979)

5.17 J.M. Liu, R. Yen, E.P. Donovan, N. Bloembergen, R.T. Hodgson, Appl. Phys. Lett. **38**, 617 (1981)

5.18 A. Garulli, M. Servidori, I. Vecchi: J. Appl. Phys. D **13**, L199 (1981)

5.19 I.W. Boyd: Appl. Phys. A **31**, 71 (1983)

5.20 K. Hoh, H. Koyama, K. Uda, Y. Miura: Jpn. J. Appl. Phys. **19**, L375 (1980)

5.21 G.G. Bentini, M. Berti, C. Cohen, A.V. Drigo, E. Jannitti, D. Pribat, J. Siejka: J. Physique **43**, C1-229 (1982)

5.22 T.E. Orlowski, H. Richter: Appl. Phys. Lett. **45**, 241 (1984)

5.23 A. Cros, R. Salvan, J. Derrien: Appl. Phys. A **28**, 241 (1982)

5.24 Y.S. Liu, S.W. Chiang, F. Bacon: Appl. Phys. Lett. **38**, 1005 (1981)

5.25 Y.S. Liu, S.W. Chiang, W. Katz: In *Laser and Electron Beam Interactions with Solids*, ed. by B.R. Appleton, G.K. Celler (Elsevier, New York 1982)

5.26 J.C.C. Tsai: In *VLSI Technology*, ed. by S.M. Sze (McGraw-Hill, London 1983)

5.27 H. Hirata, K. Hoshikawa: Jpn. J. Appl. Phys. **19**, 1573 (1980)

5.28 A.J. Schell-Sorokin, J.E. Demuth: Appl. Phys. A **39**, 13 (1986)

5.29 D.M. Zehner, C.W. White, P. Heimann, F.J. Himpsel, D.E. Eastman: Phys. Rev. B **24**, 4875 (1981)

5.30 J.E.M. Westendorp, Z. Wang, F.W. Saris: In *Laser and Electron Beam Interactions with Solids*, ed. by B.R. Appleton, G.K. Celler (North-Holland, New York 1981) p.255

5.31 E. Fogarassy, C.W. White, D.H. Lowndes, J. Narayan: In *Beam-Solid Interactions and Phase Transformations*, ed. by H. Kurz, G.L. Olson, J.M. Poate (MRS, Pittsburgh 1986) p.173

5.32 G.G. Bentini, M. Berti, A.V. Drigo, E. Jannitti, C. Cohen, J. Siejka: In *Laser Chemical Processing of Semiconductor Devices*, ed. by F.A. Houle, T.F. Deutsch, R.M. Osgood (MRS, Pittsburgh 1984) p.126

5.33 C. Cohen, J. Siejka, G.G. Bentini, M. Berti, L.F. Dona Dalle Rose, A.V. Drigo: In *Dielectric Layers in Semiconductors: Novel Technologies*, ed. by G.G. Bentini, E. Fogarassy, A. Golansk (Les Editions des Physique, Les Ulis 1986) p.197

5.34 S.W. Chiang, Y.S. Liu, R.F. Reihl: In *Laser and Electron Beam Solid Interactions and Materials Processing*, ed. by J.F. Gibbons, L.D. Hess, T.W. Sigmon (Elsevier, New York 1981)

5.35 L. Reimer: *Transmission Electron Microscopy*, 2nd ed., Springer Ser. Opt. Sci., Vol.36 (Springer, Berlin, Heidelberg 1987)

5.36 E. Fogarassy, A. Slaoui, C. Fuchs: In *Emerging Technologies for In-Situ Processing*, ed. by D.J. Ehrlich, V.T. Nguyen (Nijhoff, Dordrecht 1987)

5.37 C. W. White, D. Fathy, O.W. Holland: In *Photon, Beam and Plasma Enhanced Processing*, ed. by V.T. Nguyen, E.F. Krimmel, A. Golanski (Les Editions de Physique, Les Ulis 1987)

5.38 See, e.g., R.F. Pinizzotto: In *Ion Implantation and Ion Beam Processing of Materials*, ed. by G.K. Hubler, O.W. Holland, C.R. Clayton, C.W. White (North Holland, New York 1984) p.265

5.39 Yu.N. Molin, K.M. Salikhov, K.I. Zamaraev: *Spin Exchange*, Springer Ser. Chem. Phys., Vol.8 (Springer, Berlin, Heidelberg 1980)

5.40 K.L. Brower: Phys. Rev. B **26**, 6040 (1982)

5.41 K. Murakami, H.Itoh, K.Takita, K. Musada: Appl. Phys. Lett. **45**, 176 (1984)

5.42 H.J. Stein: J. Electrochem. Soc. **132**, 668 (1985)

5.43 D.M. Zehner, C.W. White, G.W. Ownby: Appl. Phys. Lett. **36**, 56 (1980)

5.44 P.L. Cowan, J.A. Golovchenko: J. Vac. Sci. Technol. **17**, 1197 (1980)

5.45 I.W. Boyd: In *Dielectric Layers in Semiconductors: Novel Technologies*, ed. by G.G. Bentini, E. Fogarassy, A. Golanski (Les Editions des Physique, Les Ulis 1986) p.177

Recommendations for Further Reading

Allmen M.v.: *Laser-Beam Interactions with Materials*, Springer Ser. Mat. Sci., Vol.2 (Springer, Berlin, Heidelberg 1987)

Bäuerle D.: *Chemical Processing with Lasers*, Springer Ser. Mat. Sci., Vol.1 (Springer, Berlin, Heidelberg 1986)

Biegelsen D.K., G. Rozgonyi, C.V. Shank (eds.): *Energy-Beam Solid Interactions and Transient Thermal Processing* (North-Holland, Amsterdam 1985)

Donnelly V.M., I.P. Herman, M. Hirose (eds.): *Photon, Beam and Plasma Stimulated Chemical Processes at Surfaces* (MRS, Pittsburgh 1987)

Fan J.C.C, N.M. Johnson (eds.): *Energy-Beam Solid Interactions and Transient Thermal Processing* (Elsevier, New York 1984)

Kurz H., G.L. Olson, J.M. Poate (eds.): *Beam-Solid Interactions and Phase Transformations* (MRS, Pittsburgh 1986)

Chapter 6

6.1   W. Kern, G.L. Schnable: IEEE Trans. ED-26, 647 (1979)
6.2   E.C. Douglas: Solid State Techn. **24**, 65 (1981)
6.3   A.C. Adams: In *VLSI Technology*, ed S.M. Sze (McGraw-Hill, New York 1984) p.93
6.4   J.L. Vossen, W. Kern: Physics Today 33, 26 (May 1980)
6.5   H. Sankur, J.T. Cheung: J. Vac. Sci. Technol. A 1, 1806 (1983)
6.6   M. Bass, J.R. Franchi, J. Chem. Phys. **64**, 4417 (1976)
6.7   W.M. Donnelly, M. Geva, J. Long, R.F. Karlicek: Appl. Phys. Lett. **44**, 951 (1984)
6.8   R. Solanki, W. Ritchie, G.J. Collins: Appl. Phys. Lett. **43**, 454 (1983)
6.9   H. Sankur, R.Hall: Appl. Opt. **24**, 3343 (1985) and references therein
6.10 T.F. Deutsch, D.J. Silversmith, R.W. Mountain: In *Laser Diagnostics and Photochemical,Processing for Semiconductor Devices*, ed. by R.M. Osgood, S.R.J. Brueck, H.R. Schlossberg (North-Holland, Amsterdam, 1983) p.129
6.11 D.J. Ehrlich, R.M. Osgood, T.F. Deutsch: Appl. Phys. Lett. **39**, 957 (1981)
6.12 T.Y. Sheng, L.R. Thompson, T. Hwaing, G.J. Collins: CLEO Techn. Digest (1986) p.92
6.13 Y. Hirota, O. Mikami: Electron. Lett. **21**, 77 (1985)
6.14 R. Solanki, G.J. Collins: Appl. Phys. Lett. **42**, 662 (1983)
6.15 C. Arnone, M. Rothschild, J.G. Black, D.J. Ehrlich: Appl. Phys. Lett. **48**, 1018 (1986)
6.16 See, for example, M. Hanabusa: Mat. Sci. Repts. **2**, 51 (1987) and references therein
6.17 R.F. Marks, R.A. Pollak, Ph.Avouris, C.T. Lin, Y.J. Thefaine: J. Chem. Phys. **78**, 4270 (1983) and references therein.
6.18 R.F. Marks, R.A. Pollack, Ph. Avouris: In *Laser Diagnostics and Photochemical Processing for Semiconductor Devices*, ed. by R.M. Osgood, S.R.J. Brueck, H.R. Schlossberg (North-Holland, Amsterdam 1983) p.257
6.19 R.F. Marks, R.A. Pollack: J. Chem. Phys. **81**, 1019 (1984)
6.20 B.J. Baliga, S.K. Ghandi: J. Appl. Phys. **44**, 990 (1973)
6.21 W. Peters: Technical Digest Int'l. Electron Dev. Meeting (1981) p.240

6.22 P. Dimitriou: In *Dielectric Layers in Semiconductors*, ed. by G.G. Bentini, E. Fogarassy, A. Golanski (Les Editions de Physique, Les Ulis 1986) p.349

6.23 Y. Tarui, J. Hidaka, K. Aota: Jpn. J. Appl. Phys. **23**, L827 (1984)

6.24 J.Y. Chen, R.C. Hendersen, J.T. Holland, W. Peters: J. Electrochem. Soc. **131**, 2146 (1984)

6.25 J-I Takahashi, M. Tabe: Jpn. J. Appl. Phys. **24**, 274 (1985)

6.26 Y. Mishima, M. Hirose, Y. Osaka, Y. Ashida: J. Appl. Phys. **55**, 1234 (1984)

6.27 P.K. Boyer, G.A. Roche, W.H. Ritchie, G.J. Collins: Appl. Phys. Lett. **40**, 716 (1982)

6.28 P.K. Boyer, K.A. Emery, H. Zarnani, G. J. Collins: Appl. Phys. Lett. **45**, 979 (1984)

6.29 S. Nishino, H. Honda, H. Matsunami: Jpn. J. Appl. Phys. **25**, L87 (1986)

6.30 S. Szikora, W. Krauter, D. Bäuerle: Mat. Lett. **2**, 263 (1984)

6.31 A.Sugimura, M. Hanabusa: Jpn. J. Appl. Phys. **26**, L56 (1987)

6.32 P.K. Boyer, C.A. Moore, R. Solanki, W.K. Ritchie, G.A. Roche, G.J. Collins: In *Laser Diagnostics and Photochemical,Processing for Semiconductor Devices*, ed. by R.M. Osgood, S.R.J. Brueck, H.R. Schlossberg (North-Holland, Amsterdam, 1983) p.119

6.33 R. Padmanabhan, B.J. Miller: J. Vac. Sci. Technol. A **4**, 363 (1986)

6.34 K. Tamagawa, T. Hayashi, S. Komiya: Jpn. J. Appl. Phys. **25**, L728 (1986)

6.35 H. Sankur: SPIE **459**, 78 (1984)

6.36 J.T. Yardley, A. Gupta, G. West, K.W. Beeson: In *Laser Chemical Processing of Semiconductor Devices*, ed. by F.A. Houle, T.F. Deutsch, R.M. Osgood, Extended Abstracts of Symp. B (MRS, Pittsburgh 1984) p.121

6.37 Y. Kizaki, T. Kandori, Y. Fujitani: Jpn. J. Appl. Phys. **24**, 800 (1985)

6.38 J.Y. Tsao, D.J. Ehrlich: Appl. Phys. Lett. **42**, 997 (1983)

6.39 D.J. Ehrlich, J.Y. Tsao: J. Vac. Sci. Technol. B1, 969 (1983)

6.40 D.J. Ehrlich, J.Y. Tsao: Appl. Phys. Lett. **46**, 198 (1985)

Recommendation for Further Reading

Leonberger F.J., C.H. Lee, H. Morkoc, F. Capasso (eds.): *Picosecond Electronics and Optoelectronics II*, Springer Ser. Electron. Photon., Vol.24 (Springer, Berlin, Heidelberg 1987)

Mourou G.A., D.M. Bloom, C.-H. Lee (eds.): *Picosecond Electronics and Optoelectronics*, Springer Ser. Electrophys., Vol.21 (Springer, Berlin, Heidelberg 1985)

Källbäck B., H. Beneking (eds.): *High-Speed Electronics*, Springer Ser. Electron. Photon., Vol.22 (Springer, Berlin, Heidelberg 1986)

Le Lay G., J. Derrien, N. Boccara (eds.): *Semiconductor Interfaces: Formation and Properties*, Springer Proc. Phys., Vol.22 (Springer, Berlin, Heidelberg 1987)

Vossen J.L., W. Kern: *Thin Film Processes* (Academic, New York 1978)

Chapter 7

7.1   S.S. Charschan: *Laser in Industry* (Von Nostrand, New York 1972)
7.2   W.W. Duley: CO_2 *Lasers, Effects and Applications* (Academic, New York 1976)
7.3   J.F. Ready: *Industrial Applications of Lasers* (Academic, New York 1978)
7.4   J.F. Ready (ed.): *Laser Applications in Materials Processing*, SPIE **198** (1980)
7.5   H.Koebner (ed.): *Industrial   Applications   of   Lasers* (Wiley, Chichester 1984)
7.6   J.F. Ready: *Effects of High Power Radiation*, (Academic, New York 1971)
7.7   J.F. Ready: Proc. IEEE **70**, 533 (1982)
7.8   M.I. Cohen: Bell Lab. Rec. **45**, 247 (1967)
7.9   B.A. Unger, M.I. Cohen, J.F. Milkosky: Bell Syst. Tech. J. **47**, 387 (1968)
7.10   D. Henderson, J.C. White, H.G. Craighead, I. Adesida: Appl. Phys. Lett. **46**, 900 (1985)
7.11   F.A. Houle, T.J. Chuang: J. Vac. Sci. Technol. **20**, 790 (1982)
7.12   T.J. Chuang: IBM J. Res. Develop. **26**, 145 (1982)
7.13   J.H. Brannon: In *Laser Chemical Processing of Semiconductor Devices*, Extended Abstracts of MRS Symp.B, ed. by F.A. Houle, T.F. Deutsch, R.M. Osgood (MRS, Pittsburgh 1984) p.112
7.14   G.L. Loper, M.D. Tabat: SPIE **459**, 121 (1984)
7.15   G.L. Loper, M.D. Tabat: Appl. Phys. Lett. **46**, 654 (1985)
7.16   D. Harradine, F.R. McFeely, B. Roop, J.I. Steinfeld, D. Denison, L. Hartsough, J.R. Hollahan: SPIE **270**, 52 (1981)
7.17   J.I Steinfeld, T.G. Anderson, C. Reiser, D.R. Denison, L.D. Hartsough, J.R. Hollahan: J. Electrochem. Soc. **127**, 514 (1980)
7.18   T.J. Chuang: Surf. Sci. Rept. 3, 1 (1983)
7.19   D.J. Ehrlich, R.M. Osgood, T.F. Deutsch: Appl. Phys. Lett. **38**, 1018 (1981)
7.20   S. Yokoyama, Y. Yamakage, M. Hirose: Appl. Phys. Lett. **47**, 389 (1985)

7.21 M. Hirose, S. Yokoyama, Y. Kamakage: J. Vac. Sci. Technol. B 3, 1445 (1985)

7.22 C.W. White: J. de Physique C 5, 145 (1983)

7.23 S. Unamuna, M. Toulemonde, P. Siffert: In *Chemical Processing with Lasers*, ed. by D. Bäuerle, Springer Ser. Chem. Phys., Vol.39 (Springer, Berlin, Heidelberg 1984) p.35.

7.24 R.T. Young, J. Narayan, W.H. Christie, G.A. van der Leeden, D.E. Rothe, L.J. Chen: Solid State Technol. 26, 183 (1983)

7.25 R.J. von Gutfeld: In *Laser Applications*, ed. by J.f. Ready, R.K. Erf (Academic, New York 1984) p.1

7.26 G.L. Loper, M.D. Tabat: J. Appl. Phys. 58, 3649 (1985)

7.27 M.D. Tabat: Private Communication

7.28 D.J. Ehrlich, J.Y. Tsao, C.O. Bozler: J. Vac. Sci. Technol. B3, 1 (1985)

7.29 K. Daree, W. Kaiser: Glass Technol. 18, 19 (1977)

7.30 D.L. Schaefer: U.S.Patent No. 3520687 (14 July 1970)

7.31 R.J. von Gutfeld, E.E. Tynan, R.L. Melcher, S.E. Blum: Appl. Phys. Lett. 35, 651 (1979)

7.32 R.J. von Gutfeld: Appl. Phys. Lett. 40, 352 (1982)

7.33 R.J. von Gutfeld: Denki Kagaku 52, 452 (1984)

7.34 K. Ando, N. Takeda, N. Koshizuka: Appl. Phys. Lett. 46, 1107 (1985)

7.35 J.A. Parrish, T.F. Deutsch: IEEE J. QE-20, 1386 (1984)

7.36 B.J. Garrison, S. Srinivasan: J. Appl. Phys. 57, 2909 (1985)

7.37 R. Linsker, R. Srinivasan, J.J. Wynne, D.R. Alonso: Lasers in Surgery and Medicine 4, 201 (1981)

7.38 R. Srinivasan: J. Vac. Sci. Technol. B 1, 923 (1983)

7.39 Y. Kawamura, K. Toyoda, S. Namba: Appl. Phys. Lett. 40, 374 (1982)

7.40 K. Jain, C.G. Wilson, B.J. Lin: IBM J. Res. Dev. 26, 151 (1982)

7.41 R.Srinivason, V.Mayne-Banton: Appl. Phys. Lett. 41, 576 (1982)

7.42 R. Srinivasan, R. Braren: J. Polymer Sci. 22, 2601 (1985)

7.43 J.E. Andrew, P.E. Dyer, D. Forster, P.H. Key: Appl. Phys. Lett. 43, 717 (1983)

7.44 T.F. Deutsch, M. Geis: J. Appl. Phys. 54, 1201 (1983)

7.45 T.F. Deutsch: In *Laser Processing and Diagnostics*, ed. by D. Bäuerle, Springer Ser. Chem. Phys., Vol.39 (Springer, Berlin, Heidelberg 1984) p.239

7.46 S. Rice, K. Jain: Appl. Phys. A 33, 195 (1984)

7.47 M. Latta, R. Moore, S. Rice, K. Jain: J. Appl. Phys. 56, 586 (1984)

7.48 P.E. Dyer, J. Sidhu: J. Appl. Phys. 57, 1420 (1985)

7.49 J.H. Brannon, J.R. Lankard, A.I. Baise, F. Burns, J. Kaufman: J. Appl. Phys. 58, 2036 (1985)

7.50 G. Koren, J.T.C. Yeh: Appl. Phys. Lett. 44, 1112 (1984)

7.51 G. Koren, J.T.C Yeh: J. Appl. Phys. 56, 2120 (1984)

7.52 I.W. Boyd, T.F. Boggess, A.L. Smirl, S.C. Moss: Optica Acta 33, 527 (1986)

7.53   I.W. Boyd, S.C. Moss, T.F. Boggess, A.L. Smirl: Appl. Phys.
       Lett. **45**, 80 (1984) and references therein
7.54   P.E. Dyer, R. Srinivasan: Appl. Phys. Lett. **48**, 445 (1986)
7.55   B. Danielzik, N. Fabricus, M. Rowekamp, D. von der Linde:
       Appl. Phys. Lett. **48**, 212 (1986)
7.56   R. Srinivasan, B. Braren, D.E. Seeger, R.W. Dreyfus: J.
       Macromolecules **19**, 916 (1986)
7.57   P.E. Dyer, J. Sidhu: Optics and Lasers in Engineering **6**, 67
       (1985)
7.58   R.A. Lawes: Rutherford Appleton Laboratories, UK (unpub-
       lished)
7.59   T.J. Chaung: J. Vac. Sci. Technol. **21**, 798 (1982)
7.60   T.J. Chuang: In *Laser Diagnostic and Photochemical Pro-
       cessing for Semiconductors*, ed. by R.M. Osgood, S.R.J.
       Brueck, H.R. Schlossberg (North-Holland, Amsterdam 1983)
       p.45
7.61   F.A. Houle: Appl. Phys. A **41**, 315 (1986)
7.62   K. Li, M.M. Oprysko: Appl. Phys. Lett. **46**, 997 (1985)
7.63   A. Kestenbaum, T Baer: IEEE Trans. C&HMT-3, 637 (1980)
7.64   F.W. Dabby, U.C. Paek: IEEE J. QE-8, 106 (1972)
7.65   R.G. Masters: J. Vac. Sci. Technol. A **3**, 324 (1985)
7.66   J.R. Williamson: In *Industrial Applications of High Power
       Laser Technology*, ed. by J.F. Ready, SPIE **86** (1976)
7.67   Data from "CO_2 Applications" booklet, Coherent Radiation
       Inc, Palo Alto, Calif. (1969)
7.68   L.L. Sveshnikova, V.I. Donin, S.M. Repinskii: Sov. Tech.
       Phys. Lett. **3**, 223 (1977)
7.69   I.M. Beterov, V.P. Chebotaev, N.I. Yurshina, B. Ya Yurshin:
       Sov. J. Quantum Electron. **8**, 1310 (1978)
7.70   J.E. Harry, F.W. Lunau: IEEE Trans. IA-8, 418 (1972)
7.71   J. Otto, R. Stumpe, D. Bäuerle: In *Chemical Processing
       with Lasers*, ed. by D. Bäuerle, Springer Ser. Chem. Phys.,
       Vol.39 (Springer, Berlin, Heidelberg 1984) p.320
7.72   M.W. Geis, J.N. Randall, T.F. Deutsch, P.D. Graff, K.E.
       Krohn, L.A. Stern: Appl. Phys. Lett. **43**, 74 (1983)
7.73   T.J. Chuang: J. Chem. Phys. **74**, 1461 (1981)
7.74   F.A. Houle: Chem. Phys. Lett. **95**, 5 (1983)
7.75   W. Holber, G. Rekesten, R.M. Osgood: Appl. Phys. Lett. **46**,
       201 (1985)
7.76   Y. Horiike, M. Sekine, K. Horioka, T. Arikado, M. Nakase, H.
       Okano: In *Laser Chemical Processing of Semiconductor Dev-
       ices*, Extended Abstracts, ed. by F.A. Houle, T.F. Deutsch,
       R.M. Osgood (MRS, Pittsburgh 1984) p.99
7.77   A.L. Dalisa, W.K. Zwicker, D.J. DeBitetto, P. Harnack:
       Appl. Phys. Lett. **17**, 208 (1970)
7.78   G.P. Davis, C.A. Moore, R.A. Guttscho: SPIE **459**, 115 (1984)
7.79   J.Y. Tsao, D.J. Ehrlich: Appl. Phys. Lett. **43**, 146 (1983)
7.80   W. Sesselmann, T.J. Chuang: J. Vac. Sci. Technol. B **3**, 1507
       (1985)
7.81   R.J. von Gutfeld, R.E. Acosta, L.T. Romankiw: IBM J. Res.
       Dev. **26**, 136 (1982)

7.82 G. Koren, F. Ho, J.J. Ritsko: Appl. Phys. Lett. **46**, 1006 (1985)

7.83 D.J. Ehrlich, R.M. Osgood, T.F. Deutsch: Appl. Phys. Lett. **36**, 698 (1980)

7.84 R.P. Salathe, G. Baskhara Rao: In *Laser Diagnostic and Photochemical Processing for Semiconductors*, ed. by R.M. Osgood, S.R.J. Brueck, H.R. Schlossberg (North-Holland, Amsterdam 1983) p.65.

7.85 G.C. Tisone, A.W. Johnson: Appl. Phys. Lett. **42**, 530 (1983); and in *Laser Diagnostic and Photochemical Processing for Semiconductors*, ed. by R.M. Osgood, S.R.J. Brueck, H.R. Schlossberg (North-Holland, Amsterdam 1983) p.73

7.86 M. Takai, J. Tukuda, H. Nakai, K. Gamo, S. Namba: Jpn. J. Appl. Phys. **22**, L757 (1983)

7.87 A.W. Tucker, M. Birnbaum: IEEE EDL-**4**, 39 (1983)

7.88 B. Zysset, R.P. Salathe, J.L. Martin, R. Gotthardt, F.K. Reinhart: In *Chemical Processing with Lasers*, ed. by D. Bäuerle, Springer Ser. Chem. Phys., Vol.39 (Springer, Berlin, Heidelberg 1984) p.469

7.89 P. Brewer, S. Halle, R.M. Osgood: Appl. Phys. Lett. **45**, 475 (1984)

7.90 P. Brewer, D. McClure, R.M. Osgood: In *Laser Chemical Processing of Semiconductor Devices*, Extended Abstracts, ed. by F.A. Houle, T.F. Deutsch, R.M. Osgood (MRS, Pittsburgh 1984) p.102

7.91 C.I.H. Ashby: Appl. Phys. Lett. **45**, 892 (1984)

7.92 D. Moutennet, S. Mottet, D. Riviere, J.P. Mercier: In *Chemical Processing with Lasers*, ed. by D. Bäuerle, Springer Ser. Chem. Phys., Vol.39 (Springer, Berlin, Heidelberg 1984) p.339

7.93 D.V. Podlesnik, H.H. Gilgen, P.D. Brewer, D.M. McClure, R.M. Osgood: In *Laser Chemical Processing of Semiconductor Devices*, Extended Abstracts, ed. by F.A. Houle, T.F. Deutsch, R.M. Osgood (MRS, Pittsburgh 1984) p.109

7.94 D.V. Podlesnik, H.H. Gilgen, R.M. Osgood: Appl. Phys. Lett. **45**, 563 (1984)

7.95 R.W. Hayes, G.M. Metze, V.G. Kreismanis, L.F. Eastman: Appl. Phys. Lett. **37**, 344 (1980)

7.96 R.M. Osgood, A. Sanchez-Rubio D.J. Ehrlich, V. Daneu: Appl. Phys. Lett. **40**, 391 (1982)

7.97 S. Mottet, B. Henry: Electron. Lett. **19**, 919 (1983)

7.98 V.A. Tyagai, V.A. Sterligov, G.Y. Kolbasov: Electrochim. Acta **22**, 819 (1977)

7.99 F.W. Ostermayer, P.A. Kohl: Appl. Phys. Lett. **39**, 76 (1981)

7.100 J.E. Bjorkholm, A.A. Ballman: Appl. Phys. Lett. **43**, 574 (1983)

7.101 C.I.H. Ashby, R.M. Biefeld: Appl. Phys. Lett. **47**, 62 (1985)

7.102 G. Eberhardt: In *Industrial Applications of Lasers*, ed. by H. Koebner (Wiley, New York 1984) Chap.2. p.69

7.103 M. Rothschild, C. Arnone, D.J. Ehrlich: J. Vac. Sci. Technol. B **1**, 310 (1986)

7.104 R. Solomon, L.F. Mueller: US Patent No.3,364,087 (January 1968)

7.105 B.S. Agrawalla, B.T. Dai, S.D. Allen: J. Vac. Sci. Technol. B **5**, 601 (1987)

7.106 J.G. Black, D.J. Ehrlich, M. Rothschild, S.P. Doran, J.H.C. Sedlacek: J. Vac. Sci. Technol. B **5**, 419 (1987)

7.107 M. Rothschild, J.H.C. Sedlacek, J.G. Black, D.J. Ehrlich: J. Vac. Sci. Technol. B **5**, 414 (1987)

7.108 G.V. Treyz, R. Beach, R.M. Osgood: Appl. Phys. Lett. **50**, 475 (1987)

7.109 S. Lazare, J.C. Soulignac, P. Fragnaud: Appl. Phys. Lett. **50**, 624 (1987)

Recommendation for Further Reading

Barbe D.F. (ed.): *Very Large Scale Integration VLSI*, 2nd ed., Springer Ser. Electrophys., Vol.5 (Springer, Berlin, Heidelberg 1982)

Kelly M.J., C. Weisbuch (eds.): *The Physics and Microfabrication of Microstructures and Microdevices*, Springer Proc. Phys., Vol.13 (Springer, Berlin, Heidelberg 1986)

Le Lay G., J. Derrien, N. Boccara (eds.): *Semiconductor Inter-faces: Formation and Properties*, Springer Proc. Phys., Vol.22 (Springer, Berlin, Heidelberg 1987)

Reinberg A.R.: Dry Processing for Fabrication of VLSI Devices, in *VLSI Electronics*, ed. by N.G. Einspruch (Academic, New York 1981) Vol.2

Tarui Y. (ed.): *VLSI Technology*, Springer Ser. Electrophys., Vol.12 (Springer, Berlin, Heidelberg 1986)

Chapter 8

8.1 B.M. McWilliams, I.P. Herman, F. Mitlitsky, R.A. Hyde, L.L. Wood: Appl. Phys. Lett. **43**, 946 (1983)

8.2 B.M. McWilliams: (private communication, and to be pub-lished)

8.3 G.G. Bentini, M. Berti, A.V. Drigo, E. Jannitti, C. Cohen, J. Siejka: In *Laser Chemical Processing of Semiconductor Devices*, ed. by F.A. Houle, T.F. Deutsch, R.M. Osgood (MRS, Pittsburgh 1984) p.126

8.4 E. Harari: J. Appl. Phys. **49**, 2478 (1978)

8.5 C.M. Osburn, D.W. Ormond: J. Electrochem. Soc. **119**, 591 (1972)

8.6 I.H. Pratt: Solid State Techn. **12**, 49 (December 1969)

8.7 I.W. Boyd: J. Appl. Phys. **54**, 3561 (1983)
8.8 R.R. Krchnavek, H.H. Gilgen, R.M. Osgood: J. Vac. Sci. Techn. B **2**, 641 (1984)
8.9 T.E. Orlowski, H. Richter: Appl. Phys. Lett. **45**, 241 (1984)
8.10 K. Emery, L.R. Thompson, J.J. Rocca, G.J. Collins: SPIE **459**, 82 (1984)
8.11 J. Nulman, J.P. Krusius, A. Gat: IEEE EDL-6, 205 (1985)
8.12 P.K. Boyer, G.A. Roche, W.H. Ritchie, G.J. Collins: Appl. Phys. Lett. **40**, 716 (1982)
8.13 P.K. Boyer, K.A. Emery, H. Zarnani, G. J. Collins: Appl. Phys. Lett. **45**, 979 (1984)
8.14 P.K. Boyer, C.A. Moore, R. Solanki, W.K. Ritchie, G.A. Roche, G.J. Collins: In *Laser Diagnostics and Photochemical,Processing for Semiconductor Devices*, ed. by R.M. Osgood, S.R.J. Brueck, H.R. Schlossberg (North-Holland, Amsterdam, 1983) p.119
8.15 I.W. Boyd, J.I.B. Wilson: Solid State Electr. **27**, 209 (1984)
8.16 A.M. Hodge, C. Pickering, A.J. Pidduck, R.W. Hardeman: In *Rapid Thermal Processing*, ed. by T.O. Sedgwick, T.E. Seidel, B.-Y. Tsaur (Materials Research Society, Pittsburgh 1986) p.313
8.17 C.M. Gronet, J.C. Sturm, K.E. Williams, J.F. Gibbons: In *Rapid Thermal Processing*, ed. by T.O. Sedgwick, T.E. Seidel, B.-Y. Tsaur (Materials Research Society, Pittsburgh 1986) p.305
8.18 J.C. Gelpey, P.O. Stump, R.A. Capodilupo: In *Rapid Thermal Processing*, ed. by T.O. Sedgwick, T.E. Seidel, B.-Y. Tsaur (Materials Research Society, Pittsburgh 1986) p.321
8.19 Z.A. Weinberg, T.N. Nguyen, S.A. Cohen, R. Kalish: In *Rapid Thermal Processing*, ed. by T.O. Sedgwick, T.E. Seidel, B.-Y. Tsaur (Materials Research Society, Pittsburgh 1986) p.327
8.20 J. Nulman, J.P. Krusius, N. Shah, A. Gat, A. Baldwin: J. Vac. Sci. Technol. A **4**, 1005 (1986)
8.21 N. Chan Tung, Y. Caratini, L. Liauzu: In *Dielectric Layers in Semiconductors*, ed. by G.G. Bentini, E. Fogarassy, A. Golanski (Les Editions de Physique, Les Ulis 1986) p.247
8.22 J.P. Ponpon, J.J. Grob, A. Grob, R. Stuck: J. Appl. Phys. **59**, 3921 (1986)
8.23 M.M. Moslehi, K.C. Saraswat, S.C. Shatas: Appl. Phys. Lett. **47**, 1113 (1985)
8.24 M.M. Moslehi, S.C. Shatas, K.C. Saraswat: Appl. Phys. Lett. **47**, 1353 (1985)
8.25 G.J. Fisanick, M.E. Gross, J.E. Hopkins, M.D. Fennell, K.J. Schnoes, A. Katzir: J. Appl. Phys. **57**, 1139 (1985)
8.26 M.E. Gross, A. Appelbaum, P.K. Gallagher: Appl. Phys. Lett. **61**, 1628 (1987)
8.27 J. Bohandy, B.F. Kim, F.J. Adrian: J. Appl. Phys. **60**, 1538 (1986)
8.28 H. Sankur, J.T. Cheung: J. Vac. Sci. Technol. A1, 1806 (1983)

8.29 D.J. Ehrlich, R.M. Osgood, D.J. Silversmith, T.F. Deutsch:
      IEEE EDL-1, 101 (1980)
8.30 Y. Rytz-Froideveau, R.P. Salathé, H.H.Gilgen: Appl. Phys. A
      37, 121 (1985)
8.31 J.A. Yasaitis, G.H. Chapman, J.I. Raffel: IEEE EDL-3, 184
      (1982)
8.32 J.Y. Tsao, D.J. Ehrlich, D.J. Silversmith, R.W. Mountain:
      IEEE EDL-3, 164 (1982)
8.33 J.G. Black, D.J. Ehrlich, M. Rothschild, S.P. Doran, J.H.C.
      Sedlacek: J. Vac. Sci. Technol. B 5, 419 (1987)
8.34 D.J. Ehrlich, J.Y. Tsao, D.J. Silversmith, J.H. Sedlacek,
      R.W. Mountain, W.S. Graber: IEEE EDL-3, 164 (1984)
8.35 B.M. McWilliams, H.W. Chin, I.P. Herman, R.A. Hyde, F. Mit-
      litsky, J.C. Whitehead, L.L. Wood: SPIE **459**, 49 (1984)
8.36 H. Zarnani, P.K. Boyer, M. Fathipour, C.W. Wilmsen, R. Sol-
      anki, G.J. Collins: In *Laser Chemical Processing of Semi-
      conductor Devices*, ed. by F.A. Houle, T.F. Deutsch, R.M.
      Osgood (MRS, Pittsburgh 1984) p.122
8.37 B.E. Deal, A.S. Grove: J. Appl. Phys. **36**, 3770 (1965)
8.38 I.W. Boyd, J.I.B. Wilson: J. Appl. Phys. **53**, 4166 (1982)
8.39 I.W. Boyd, J.I.B. Wilson, J.L. West: Thin Solid Films **83**,
      L173 (1981)
8.40 S.A. Schafer, S.A. Lyon: J. Vac. Sci. Technol. **19**, 494
      (1981)
8.41 C. Fiori: Phys. Rev. Lett. **52**, 2077 (1984)
8.42 Y. Mishima, M. Hirose, Y. Osaka, Y. Ashida: J. Appl. Phys.
      55, 1234 (1984)
8.43 A. Cros, R. Salvan, J. Derrien: Appl. Phys. A **28**, 241
      (1982)
8.44 Y.S. Liu, S.W. Chiang, F. Bacon: Appl. Phys. Lett. **38**, 1005
      (1981)
8.45 E. Fogarassy, C.W. White, D.H. Lowndes, J. Narayan: In *Beam
      Solid Interactions and Phase Transformations*, ed. by H.
      Kurz, G,L. Olson, J.M. Poate (MRS, Pittsburgh 1986) p. 173

Recommendation for Further Reading

Allmen, M.von: *Laser-Beam Interactions with Materials*, Springer
Ser. Mat. Sci., Vol.2 (Springer, Berlin, Heidelberg 1987)

Aussenegg, F.R., A. Leitner, M. Lippitsch (eds.): *Surface Studies
with Lasers*, Springer Ser. Chem. Phys., Vol.33 (Springer,
Berlin, Heidelberg 1983)

Bäuerle, D.: *Chemical Processing with Lasers*, Springer Ser. Mat.
Sci., Vol.1 (Springer, Berlin, Heidelberg 1986)

Donnelly V.M., I.P. Herman, M. Hirose (eds.): *Photon Beam and
Plasma Stimulated Chemical Processes at Surfaces*, MRS Symp.
Proc. Vol.75 (MRS, Pittsburgh 1987)

Gibbons, J.F.(ed.): *CW Beam Processing of Silicon and Other Semiconductors*, Vol.17 (Academic, New York 1984)

Gutfeld R.J.von, J.E. Greene, H. Schlossberg (eds.): *Beam Induced Chemical Processes*, Extended Abstracts Symp.D, 1985 Fall Meeting Materials Research Society, Boston (MRS, Pittsburgh 1985)

Howard R.E., E.L. Hu, S. Namba, S.W. Pang (eds.): *Science and Technology of Microfabrication*, MRS Symp Proc. Vol.76 (MRS, Pittsburgh 1987)

Kurz H., G.L. Olson, J.M. Poate (eds.): *Beam-Solid Interactions and Phase Transformations*, MRS Symposia Proc. Vol.55 (MRS, Pittsburgh 1986)

Le Lay G., J. Derrien, N. Boccara (eds.): *Semiconductor Interfaces: Formation and Properties*, Springer Proc. Phys., Vol.22 (Springer, Berlin, Heidelberg 1987)

Nguyen V.T., D.J. Ehrlich: *Emerging Technologies for In-Situ Processing* (Nijhoff, Dordrecht 1987)

Wood R.F., C.W. White, R.T. Young (eds.): *Pulsed Laser Processing of Semiconductors*, Semiconductors and Semimetals, Vol.23 (Academic, New York 1984)

Subject Index

Phase changes, *see* Laser induced heating

Phonons 46,64, *see also* Optical absorption of materials

Photochemical processing 3,61,91, 116,214,219,226

Photodesorption 52

Photodissociation 20,165,244,259

Photoenhanced growth 158,164,183

Photoresist 241,254,262

   contrast factor 109

Photosensitisation 224,226,230, 243

Photothermal processing 90,148, 157,188,213,227,230

Physisorption 49, *see also* Adsorption

Plasma 48,221

Plasma frequency 26

PMMA 63,233,241,255

Polyimide 241,255,260

Polymerisation 63

Powder formation 231

Precursors 25,280

Predissociation 24

Profile measurements, *see* Laser beams

Publications 3

Rapid thermal oxidation 171,179, 278

Reaction rates 89

   CW laser controlled rates 97

   high laser intensity levels 93

   low laser intensity levels 90

Reflectivity 29,37,40, *see also* Optical absorption of materials

Repair 284

Resistors 267

Ripples 126

SF_6 21

Si 34,35,38,40,46,65,71,76

Si_3N_4 31,135,144,160,184,204,278, 284

   laser CVD 229

   photosensitised deposition 230

   plasma deposition 229

   powder 231

Silicides 97

SiO 169

SiO_2 21,31,36,154,278

   deposition 148,211,223

   etching 240,241

   growth models 143

   growth rate 137,139

   infrared properties 170,173

   laser assisted growth 157

   native layer 139

   optical properties 36

   photoenhanced growth 158,164

   physical properties 154

   RHEED measurements 194

   step coverage 226

   structure 175

Slip lines 160,171

Spatial filtering, *see* Laser beams

Spatial resolution 107

Stranski-Krastanov mechanism 85

Supercooling 83

Surface cleaning 51,63,206